Biodiesel

Biodiesel

Growing a New Energy Economy

GREG PAHL

Foreword by Bill McKibben

CHELSEA GREEN PUBLISHING COMPANY
WHITE RIVER JUNCTION, VERMONT

Managing Editor: Collette Leonard
Project Editor: Marcy Brant
Developmental Editor: Ben Watson
Copy Editor: Nancy Ringer
Proofreader: Rachael Cohen
Indexer: Peggy Holloway
Designer: Peter Holm, Sterling Hill Productions
Design Assistant: Daria Hoak, Sterling Hill Productions

Printed in Canada
First printing, January 2005

Library of Congress Cataloging-in-Publication Data

Pahl, Greg.
Biodiesel : growing a new energy economy / Greg Pahl.
p. cm.
Includes bibliographical references and index.
ISBN 1-931498-65-2 (pbk.)
1. Biodiesel fuels. I. Title.
TP359.B46P34 2005
662'.669—dc22

2004020429

Chelsea Green Publishing Company
Post Office Box 428
White River Junction, VT 05001
(800) 639-4099
www.chelseagreen.com

This book is dedicated to the many people around the world who have worked tirelessly to make biodiesel a reality.

"The use of vegetable oils for engine fuels may seem insignificant today, but such oils may become, in the course of time, as important as petroleum and the coal-tar products of the present time."

RUDOLF DIESEL, 1912

CONTENTS

FIGURES

ACKNOWLEDGMENTS

While I worked on this project, I met and spoke with some wonderful, enthusiastic people who are committed to helping the global community free itself from its dependency on fossil fuels.

I would like to acknowledge the many people who were so generous with their time and advice. This book would not have been possible without people like Dr. Charles Peterson, University of Idaho, Moscow, Idaho; Dr. Thomas Reed, Golden, Colorado; Professor Leon Schumacher, University of Missouri at Columbia; Professor Jon Van Gerpen, Iowa State University, Ames, Iowa; Bill Ayres, Kansas City, Kansas; Bob King, president, and Daryl Reece, vice president, Pacific Biodiesel Inc., Kahului, Hawaii; Jerrel Branson, president, Best BioFuels, LLC, Austin, Texas; Tom Leue, president, Homestead Inc., Williamsburg, Massachusetts; John Hurley, Dog River Alternative Fuels, Berlin, Vermont; Greg Liebert, president, Vermont's Alternative Energy Corporation, Williston, Vermont; Gene Gebolys, president, World Energy Alternatives LLC, Chelsea, Massachusetts; Dennis Griffin, chairman, Griffin Industries, Cold Spring, Kentucky; Gary Haer, West Central Coop, Ralston, Iowa; Jeff Probst, president and CEO, Blue Sun Biodiesel, Fort Collins, Colorado; Bob Clark, sales manager, biodiesel division, Imperial Western Products, Coachella, California; Joe Jobe, executive director, Charles Hatcher, former regulatory director, and Jenna Higgins, director of communications, the National Biodiesel Board, Jefferson City, Missouri; Neil Caskey, special assistant to the CEO, American Soybean Association, Saint Louis, Missouri; Gerhard H. Knothe, National Center for Agricultural Utilization Research, Peoria, Illinois; Alan Weber, MARC-IV Consulting, Columbia, Missouri; John Van de Vaarst, deputy area director, Beltsville Agricultural Research Center, Beltsville, Maryland; Nicole Cousino, San Francisco, California; Sarah Lewison, San Francisco, California; Maria "Mark" Alovert, San Francisco, California; Professor Phanindra Wunnava, Middlebury College, Middlebury, Vermont; Kyoko Davis, Middlebury College, Middlebury, Vermont; Terry Mason, North Wolcott, Vermont; Raffaello Garofalo, secretary general, European

Biodiesel Board, Brussels, Belgium; Werner Körbitz, chairman, Austrian Biofuels Institute, Graben, Austria; Manfred Wörgetter, deputy director, head of Research Agricultural Engineering, Federal Institute for Agricultural Engineering, Wieselburg, Austria; Dr. Martin Mittelbach, Institute of Organic Chemistry, University of Graz, Graz, Austria; Lourens du Plessis, Bio/Chemtek division, CSIR, Pretoria, South Africa; and Darryl Melrose, Biodiesel SA, Merrivale, South Africa.

I offer my sincere thanks to all the wonderful folks at Chelsea Green Publishing who helped bring this book to completion. In particular I want to thank Ben Watson and Marcy Brant, my editors, who helped guide me through the process. And I especially want to thank Margo Baldwin for her unwavering enthusiasm and support for this project.

I also want to thank the Austrian Biofuels Institute, the Union for the Promotion of Oil and Protein Plants (Germany), and the National Biodiesel Board, who supplied the information for the charts, and anyone else I may have forgotten to mention. All of your contributions, both large and small, are greatly appreciated.

Last, but by no means least, I want to thank my wife, Joy, for her help in chasing down obscure facts, proofreading, making suggestions, and generally putting up with me while I was trying to meet deadlines.

GREG PAHL
JUNE 2004

FOREWORD

For a hundred years we've powered our lives the easiest possible way—by tapping into those pools of hydrocarbons left by the eons. That energy is highly concentrated, easily portable. It was just waiting there, for us to come along and scoop up, and so we did.

Now we've begun to realize that petroleum can't be what we use to power our next century. Not only is the supply starting to dwindle, and hence get more expensive to extract (both in terms of money and blood), but it's also become clear just what a high environmental price we've paid for the convenience. If nothing else, the news that our planet is likely to warm five degrees this century should be enough to set us looking for new paths.

Biodiesel is one of the most intriguing of those new possibilities. For ancient biology, compressed by the weight of time into petroleum, it substitutes present-day biology: crops of soybeans and rapeseed and maybe even algae, grown by present-day farmers, processed into a diesel fuel substitute that works just fine in modern Volkswagens and Mack trucks and school buses—even in the oil-burning furnace down in the basement. It is potentially a truly sweet solution, offering a new market for hard-pressed local farmers even as it begins to help solve some of our most pressing environmental problems. Greg Pahl's book, though it is impeccably careful and well-documented, nonetheless brims over with a justified excitement at the possibility of this homegrown energy.

It also manages to raise the right questions (and raise them early enough) so that we can perhaps build a structure for this developing industry that serves local farmers and processors instead of simply corporate agribusiness giants: since this project is largely dependent on public funding for a jumpstart, that is not too much to ask. The proper scale is a key question—clearly, though, it's somewhere between the guy in his garage brewing old fryer oil into fuel and the Cargills and Archer Daniels Midlands of the world simply adding energy to their portfolios. By tempering his enthusiasm with reality on these questions, Pahl does an enormous service to the future.

He's also realistic about an important fact: biodiesel is not going to solve our energy and environment woes by itself. It might replace 10 or 20 percent of our current diesel fuel use. That's good, but it's not a silver bullet against global warming. There *are* no silver bullets—every solution, from new lightbulbs to windmills to solar rooftops to higher mileage standards to biodiesel is going to get us a few percentage points of the way to where we need to go. Energy of the future will be far more diffuse, and harder to gather, than the current concentrated pools of oil. It's crucial that we recognize that fact and its key implication—that every ounce of effort put into new fuel supplies must be matched by an equal attention to conservation, to learning to live elegantly with less. This is completely possible—Europe, whose efforts on biodiesel Pahl chronicles in comprehensive fashion—manages to use about half as much energy per capita overall. And yet Europeans lead lives of civilized dignity.

This book excited me enormously. I can imagine the day when the schoolbuses on the rural rounds in my county run on the oilseed crops that their passengers can see out the window; when the ferries across Lake Champlain give off that slight french-fry whiff as they ply the waters; when the dairy farmers who are going broke raising milk have something else to grow. And when we can hold our heads a little higher, realizing that we're taking new responsibility for the energy we use. Pahl is a visionary, but a visionary with his feet firmly planted in the soil. May his vision flower, and soon!

Bill McKibben

INTRODUCTION

We are running out of oil. This is an undeniable fact. The only remaining question is not *if* but *when*.

However, pumping the global oil barrel dry is not the immediate problem. The more imminent danger is what happens when demand outstrips supply: dramatic price increases for oil, followed by exponential price increases. Recent record-high gasoline prices of well over $2.00 per gallon in the United States are only a hint of what is coming in the near future. The main problem is that, even now, annual demand for oil is four times greater than the volume of new oil reserves discovered (which peaked back in the late 1960s and early 1970s). Moreover, many of the highly publicized "huge" new recent oil finds will add only a few *days'* supply to the global oil market, which currently consumes 81 million barrels a day. By 2025, that demand is expected to climb to around 121 million barrels a day, according to the U.S. Energy Information Administration. What's more, as demand continues to increase, production from most of the largest existing oil fields is declining at about 4 to 5 percent annually, and world production of oil is expected to peak around 2010. Sooner or later, the line on the chart for demand that's heading up will cross the line on the chart for supply that's coming down. When that occurs, we will have reached the critical "tipping point."

The utter chaos this would cause in the global economy is almost too frightening to contemplate. But we'd better take a good, close look, because the date for this scenario will arrive long before we actually run out of oil. And this date is coming much sooner than most people realize. The experts, as always, are divided on when it will take place. One of the most optimistic views, promoted by the U.S. Department of Energy, maintains that oil production won't peak until 2037. Many observers, though, feel this estimate is far too optimistic, especially considering the huge increase in demand from countries like China, which overtook Japan as the world's second-largest oil consumer in 2003. Renowned petroleum geologist Colin Campbell estimates that global extraction of oil will peak before 2010. Geophysicist Kenneth Deffeyes says the date for maximum production was 2004.[1]

While these predictions may sound alarmist, the recent massive accounting scandal involving oil giant Royal Dutch/Shell and the subsequent 22 percent (4.35 *billion* barrels) cut in the company's petroleum reserve estimates is viewed by some industry experts as just the tip of the iceberg of overinflated reserve figures. If these calculations are as misleading as some people suspect, then we are in for a very rough ride in the very near future. Regardless of whoever turns out to be right about the timing of oil's tipping point, most middle-aged people probably will live to see the consequences. They may wish they hadn't. And as for the younger generations, well, they just may be out of luck.

Nearly two-thirds of the world's proven oil reserves are located in the eleven countries that make up the Organization of the Petroleum Exporting Countries (OPEC): Algeria, Indonesia, Iran, Iraq, Kuwait, Libya, Nigeria, Qatar, Saudi Arabia, the United Arab Emirates, and Venezuela. The fact that most of these nations are located in the increasingly unstable Middle East is not especially reassuring. And the fact that the former colonial powers of Europe, and more recently the United States, have been involved in numerous wars in the region, in an ongoing attempt to protect the uninterrupted flow of oil, is even more troubling and does not augur well for the future. Figure 1, a chart from the U.S. Department of Energy's Office of Transportation Technology, shows where most of the world's remaining reserves are located.

Most of the global economy is so dependent on the price of oil that any substantial price increase means big trouble. The huge price increases for oil predicted for the time after we reach the tipping point will lead unquestionably to much higher food prices, as well as to disruptions in the increasingly globalized system of food production and distribution. As a result, millions of people, particularly in struggling Third World countries, almost certainly will starve. Prices for most other goods also will rise dramatically. A severe global economic depression, massive unemployment, political instability, and more international conflict are almost certain to follow. It's conceivable that industrialized society as we know it could col-

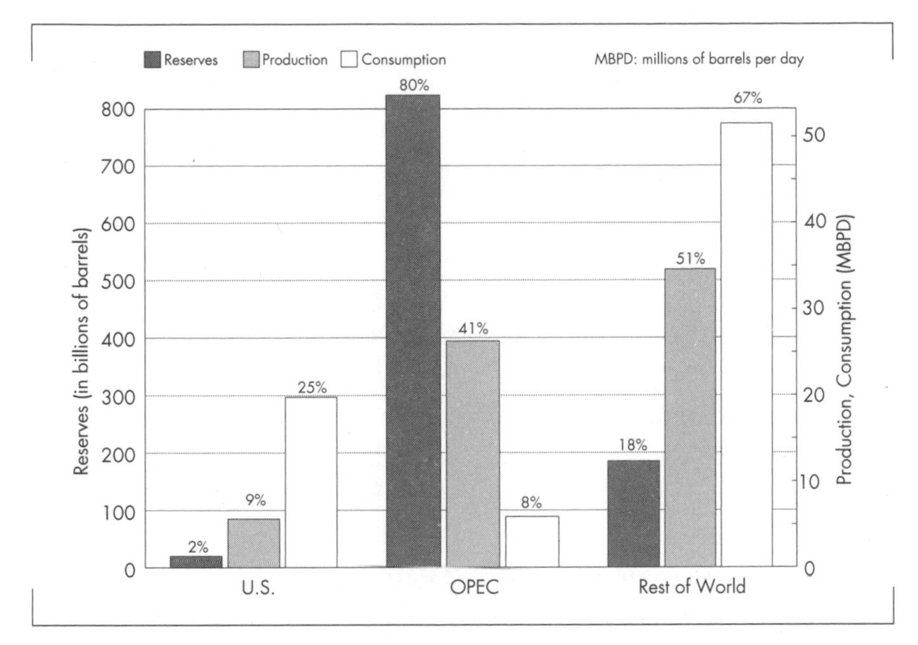

Figure 1. World Oil Reserves, Production, and Consumption, 2002 (U.S. Department of Energy)

lapse—even before we are decimated by the consequences of global warming caused by the use of fossil fuels.

This second scenario, which environmental activists and scientists have been warning about for many years, is beginning to look increasingly likely. The five warmest years in recorded weather history have taken place in the past six years, according to the World Resources Institute. And it's not just environmentalists who are worried. In February 2004, a secret Pentagon study that the Bush Administration had tried to suppress was leaked to the press. The study warned about the possible consequences of sudden climate changes caused by global warming and offered a terrifying picture of a global catastrophe costing millions of lives due to wars and natural disasters. The threat to global stability posed by global warming far surpasses that of terrorism, according to the study. Will it be a global economic meltdown caused by oil prices or one caused by global warming? Either way, we're toast.

Is there any hope of avoiding this terrible scenario? Perhaps, but time is rapidly running out. In fact, it may have already run out. If so, we won't know for sure until it's too late to do anything about it. Ominously, a number of nations that have worked tirelessly to get the United States to sign the 1997 Kyoto agreement to cut greenhouse gas emissions recently have begun to suggest that, in the absence of meaningful progress, the world should "prepare for the worst." Unfortunately, many politicians—especially in the United States—are still in denial and have long resisted doing anything about cutting greenhouse gas emissions, saying that it "would damage the U.S. economy" or "cost jobs." What these head-in-the-sand politicians fail to grasp is that if the worst-case scenarios about global warming come to pass, the catastrophic economic damage caused will be far worse than any possible costs of trying to meet the targets of the Kyoto accord. This kind of shortsighted stupidity imperils all of us, and it is particularly galling to the millions of concerned citizens around the world (and even in the United States) who rightly view the United States as the world's largest consumer of energy—and the biggest polluter. This country, with just 4.5 percent of the global population, consumes about 25 percent of the world's energy and releases roughly 25 percent of the global carbon dioxide emissions, according to the Energy Information Administration. Without active, constructive participation by the United States to help deal with these critical issues, the rest of the global community is effectively stymied.

It's going to take a huge cooperative effort on the part of the *entire* global community to wean ourselves from our present addiction to fossil fuels in general, and petroleum in particular. Many alternative strategies that rely on various forms of renewable energy, including wind, solar, and geothermal, are gaining in popularity. But running most vehicles directly on these forms of renewable energy, given present technology, is not practical. "And even though technology allows for greater fuel efficiency than ever before, cars and other forms of transportation account for nearly 30 percent of world energy use and 95 percent of global oil consumption," according to the Worldwatch Institute's recent annual report, *State of the World 2004*.

Ninety-five percent of global oil is consumed for transportation! This statistic points right to the heart of the problem. Some people suggest that compressed natural gas (CNG) could serve as a substitute for oil. But using CNG would, at the very least, require expensive retrofitting of vehicles. Unfortunately, natural gas, though cleaner burning, is still a fossil fuel, and natural gas prices have been soaring while world reserves are shrinking almost as fast as those for oil. Hydrogen-powered fuel cells are widely viewed as the ultimate solution for the transportation sector. But hydrogen is produced by the electrolysis of water, and the electricity required to produce enough hydrogen to fuel all the cars in the United States would require four times the present capacity of the national grid (unfortunately, the present grid relies on nonrenewable energy sources for 91 percent of its capacity). What's more, there is no infrastructure for the production and delivery of the vast amounts of hydrogen that would be required. The transition to hydrogen is, at best, a long, long way off.

In the meantime, there is one liquid fuel that is both renewable and can be used in a wide range of vehicles without any modifications to the engines. That fuel is biodiesel. For many years, farmers, environmentalists, and renewable energy advocates in Europe and the United States have been promoting the use of biodiesel as an alternative to at least a portion of the petroleum-based diesel fuel market. But it wasn't until the attacks on the World Trade Center on September 11, 2001, that most Americans finally began to realize the implications of their overreliance on oil—especially Middle Eastern oil—and its heavy economic, political, social, and military costs. The U.S.-led military campaign in Afghanistan and the subsequent ill-advised invasion of Iraq, with its terrible and costly aftermath, have added urgency to the movement seeking to wean the United States from its almost total addiction to petroleum-based fuels. While many strategies are currently being pursued to accomplish that end, biodiesel is one of the most intriguing and, until fairly recently, one of the least publicized in the United States.

In most of Europe, the general public is aware of biodiesel due to strong governmental support, but the sudden emergence of biodiesel from relative obscurity in the United States has taken many Americans by surprise. While other renewable energy strategies such as solar, wind, ethanol, and fuel cells have received most of the media attention, a group of Midwestern soybean farmers and other entrepreneurs have been quietly building biodiesel production capacity and infrastructure. At the same time, a number of federal and state agencies and independent organizations have been testing and evaluating biodiesel performance and setting up fuel production standards, laying the foundation for a new sustainable energy industry. Based on that firm foundation, the biodiesel industry is beginning to experience dramatic growth, both in production capacity and in the number of retail fuel outlets across the country. Despite that growth, however, many people still have only a vague idea of what biodiesel is, and fewer still understand that it can be used for more than fueling diesel-powered cars or pickup trucks.

What, exactly, is biodiesel, and why is it generating so much excitement? First, it's important to understand that even though *diesel* is part of its name, pure biodiesel does not contain any petroleum diesel or fossil fuel of any sort. Biodiesel generally falls under the category of *biomass*, which refers to renewable organic matter such as energy crops, crop residues, wood, municipal and animal wastes, et cetera, that are used to produce energy. More specifically, *biofuels*, a subcategory of biomass, includes three energy-crop-derived liquid fuels: ethanol (usually referred to as grain alcohol), methanol (usually referred to as wood alcohol), and biodiesel. Technically a fatty acid alkyl ester, biodiesel can be easily made through a simple chemical process from virtually any vegetable oil, including (but not limited to) soy, corn, rapeseed (canola), cottonseed, peanut, sunflower, avocado, and mustard seed. But biodiesel can also be made from recycled cooking oil or animal fats. There have even been some promising experiments with the use of algae as a biodiesel feedstock. And the process is so simple that biodiesel can be made by virtually anyone, although the chemicals required (usually lye and methanol) are hazardous and need to be handled with extreme caution.

Best of all, biodiesel feedstock sources are renewable and can be produced locally. While fossil fuels were formed over millions of years (and are being rapidly depleted), biodiesel can be created in just a few months. The source of the energy content in biodiesel is solar energy captured by feedstock plants during the process of photosynthesis, inspiring some to refer to the fuel as "liquid solar energy." And the plants grown to make more biodiesel naturally balance most of the carbon dioxide emissions created when the fuel is combusted, eliminating a major contributing factor to global warming. What's more, the resulting fuel is far less polluting than its petroleum-based alternative; biodiesel produces lower quantities of cancer-causing particulate emissions, is more biodegradable than sugar, and is less toxic than table salt. And because it can be produced from domestic feedstocks, biodiesel reduces the need for foreign imports of oil, while simultaneously boosting the local economy. No wonder there is so much enthusiasm, especially in the agricultural community, about biodiesel: farmers can literally grow their own fuel.

While biodiesel may be a relative newcomer to the United States, in Europe it has enjoyed widespread acceptance as a vehicle fuel (as well as a heating fuel in some countries) due to deliberate government tax policies that favor its use. In Germany, for example, where diesel engines power close to 40 percent of passenger cars, more than 1,800 filling stations offer biodiesel at a price competitive with that of regular diesel due to large tax breaks and subsidies for alternative fuels. (In the United States, by comparison, where only about 1 percent of automobiles are diesel powered and tax policies are generally not as favorable, the number of gas stations offering biodiesel is just over 300.) The expansion of the European Union in May 2004 offered a good deal of additional potential for continued growth in biodiesel production and use in the so-called new accession nations. There is good potential for the industry in many other countries around the world as well.

This book describes biodiesel's dramatic growth and its potential to help pave the way for an eventual transition from fossil fuels to a wide range of renewable energy sources. It is divided into four parts. In part one we begin this exploration with a look at biodiesel basics. We'll travel back in time to the nineteenth century to discover the roots of the device that has made the whole biodiesel movement possible—the diesel engine—and we'll learn about its tireless inventor, Rudolf Diesel, and his renewable-fuel vision that is only now being realized. Then we'll fast-forward to the 1970s to see how and why biodiesel was developed. We'll also go through the biodiesel production process and examine the many (sometimes quirky) raw materials from which biodiesel can be made, and we'll explore the fuel's environmental impact. Finally, we'll focus on the modern diesel engine and the many uses of biodiesel fuels.

In part two, we'll travel to Europe, the leader in global biodiesel production, to see why Germany, France, and Italy combined produce nearly eighteen times more biodiesel than the entire United States, and how they managed to gain such a decisive lead. We'll also visit the other European nations that are busily expanding their biodiesel industries and check out some of the more interesting developments there as well. Then we'll travel around the world to see other exciting new biodiesel projects from India to Australia and Japan to Brazil.

In part three, our biodiesel odyssey finally arrives in the United States. We'll look at some early biofuels projects by Henry Ford and see how the fledgling biofuels industry was eliminated by Big Oil and other business pressures. Then we'll follow the revival of biofuels after the 1973 OPEC oil crisis and the subsequent development of the biodiesel movement in the United States. We'll also learn about some of the main players in today's biodiesel industry and take a look at the complex world of biodiesel politics. Then we'll hear from a number of high-profile celebrities about their use of biodiesel and also see what many different people all across the country are doing with this renewable biofuel today.

In part four, we take stock of the current state of the industry and explore some of the key issues that need to be confronted if it is going to be successful. We'll also look briefly at a number of concerns that some observers have raised about the ongoing dramatic growth of the industry. And finally, we'll look into a crystal ball with some of the industry's key players to try to envision the outlines of where the biodiesel industry may be headed in the future.

While biodiesel is not the single solution to all our energy problems, it can be part of the transition from our current near-total dependency on fossil fuels, while at the same time creating jobs, assisting farmers, reducing harmful emissions, and promoting greater energy security. Biodiesel, along with a wide range of other renewable energy strategies, coupled with dramatically increased energy efficiencies, should be able to meet our energy needs well into the future. However, in order to achieve that goal, we need to rapidly increase the pace of the transition to a new energy economy today, while there's still time.

A few technical notes. The word *diesel* is used throughout this book but has different meanings in different contexts. In an attempt to clarify this, the word *diesel* (in lowercase) refers to the engine or the industry. The capitalized word *Diesel* refers to the engine's inventor, Rudolf Diesel, or his family. The word *petrodiesel* refers to the petroleum-based diesel fuel that has been used to run diesel engines since the early 1900s. And, finally, *biodiesel* refers to the renewable biofuel that is the main focus of this book.

Also, the tangle of various units of measure used in different countries around the world was a particular challenge. Metric and U.S. equivalents for various units of measure are provided here and there to help readers in most countries make sense of the statistics cited. But in an attempt to avoid overkill, not every single unit of measure has been converted. Unless otherwise noted, *tons* refer to metric tons and *gallons* refer to U.S. gallons.

PART ONE

Biodiesel Basics

1

Rudolf Diesel

The death of Rudolf Diesel is shrouded in mystery. Diesel, the inventor of the engine that now bears his name, was sailing from Antwerp, Belgium, to Harwich, England, to attend the annual directors' meeting of the British Diesel Company and the groundbreaking ceremonies for a new diesel engine plant at Ipswich when he disappeared. On the evening of September 29, 1913, Diesel embarked on the steamship *Dresden*, accompanied by fellow director George Carels and Alfred Luckmann, the chief construction engineer of the diesel engine company.

After the ship departed from Antwerp, the three men had dinner, followed by a leisurely stroll on deck, where they talked for some time. "Dr. Diesel was in the very best spirits. The conversation was cheery and buoyant," Carels is quoted as saying in a September 30, 1913, *New York Times* article about the incident. Around 10 p.m., Diesel and his companions decided to turn in for the night, and they descended to their cabins. Diesel stopped briefly in his cabin but then came down the corridor to shake Carels's hand and wish him good night. "I will see you tomorrow morning," Diesel said to Carels.

But the next morning, when Diesel failed to show up for breakfast, Carels and Luckmann went to his cabin to find him. They knocked on the door, and when there was no response, they looked inside. The cabin was empty. A quick search of the cabin revealed that Diesel's bed had not been slept in, although his nightshirt was neatly laid out and his watch had been left where he could see it from the bed. Nothing appeared to have been disturbed. The *Dresden*'s crew was immediately notified and a quick search of the ship was conducted, but Diesel could not be found. The missing man's hat and overcoat, however, were discovered neatly folded beneath the afterdeck railing. After the ship arrived in Harwich, the German vice-consul was notified, and he instigated a thorough search of the entire ship—to no avail. Diesel had completely vanished. "As his landing ticket had not been given up, we felt certain that Dr. Diesel could not have landed, and, as he was not to be found on board, we could not think otherwise than that he disappeared overboard in the course of the night," Carels said. But how? And why? Was it an accident? Had he committed suicide? Or had he, perhaps, been murdered?

There did not appear to be any logical explanation for Diesel's disappearance. The sea had been relatively calm (for the English Channel) on the night in question, eliminating the possibility that Diesel had been accidentally swept overboard during a storm. The press had a field day. Headlines such as "Diesel Murdered by Agents from Big Oil Trusts" or "British Secret Service Eliminates Diesel" quickly appeared in European newspapers. In the United States, "Creator of the Diesel Engine Executed as a Traitor to Secure U-boat Secrets" and "Inventor Thrown into the Sea to Stop Sale of Patents to British Government" screamed from page one. It was also suggested by some that "embittered enemies" or "business competitors" of Diesel might have arranged his death. There was at least some credibility to the suspicions raised about the U-boat issue, since Diesel had already convinced the French Navy in 1904 to install his engine in a number of their submarines. And one of his less publicized reasons for traveling to England was to try to do the same for the British Navy.

On October 10, 1913, the Belgian pilot steamer *Coertsen* spotted a body floating in the North Sea. Despite stormy weather, a small lifeboat was lowered into the water and the body was taken on board long enough to recover a few personal effects from its clothing. Following the normal custom of the time, the badly decomposed body was quickly returned to the sea. The personal items were brought to the Dutch port of Vlissingen, where they were subsequently identified by Diesel's son, Eugen, as having belonged to his father.

The exact circumstances surrounding Diesel's death will never be known. There was no official investigation or trial, not even a hearing by the ship's company. Since the body was never recovered again, there was no autopsy and no official coroner's report. Diesel did not leave a suicide note. He did not leave a will. And he apparently never mentioned any intention to kill himself. He did, however, leave an enigmatic cross penciled in his notebook after the date for September 29.

Shortly after Rudolf Diesel disappeared, his wife, Martha, opened a bag that her husband had given to her just before his ill-fated voyage, with directions that it should not be opened until the following week. She discovered twenty thousand German marks in cash and a number of financial statements indicating that their bank accounts were virtually empty. At the time, the general public had assumed that the Diesel family was enormously wealthy, but in fact they were almost broke.

His Early Years

Rudolf Christian Karl Diesel was born in Paris on March 18, 1858. His parents, Theodor and Elise Diesel, were both originally from the Bavarian city of Augsburg. Rudolf had an older sister, Louise, born in 1856, and a younger one, Emma, born in 1859. Theodor, a leather worker who had emigrated from Germany around 1850, ran a small leather shop in Paris. Rudolf was a shy but intelligent child who spent a lot of time drawing. Early on, Rudolf exhibited an interest in things mechanical. He dismantled the family's cherished cuckoo clock but was embarrassed when he was unable to reassemble it.[1] As he grew older, Rudolf excelled in his studies, learned to speak three languages—German, French, and English—and spent much of his spare time in the Conservatoire des Arts et Metiers, the oldest technical museum in Paris, which housed an interesting collection of mechanical inventions. Rudolf completed his elementary schooling and was awarded a bronze medal for academic excellence. His future looked extremely promising.

Then fate stepped in and severely disrupted the family's normal routine. On July 19, 1870, France declared war on Prussia. Ironically, from the start the war did not go well for the overconfident French, and as one humiliating military defeat followed another, waves of refugees began to flow into Paris. The Diesels (and anyone else of German descent) were increasingly viewed with suspicion, and they were soon expelled from France. On September 6 the family boarded a steamer from Dieppe to England. A few days later the refugees arrived in London, where, after much difficulty, Theodor Diesel finally managed to find a low-paying job. Rudolf, now twelve years old, was able to resume his studies in a London school. In his spare time he visited the British Museum and the South Kensington Museum, where he often lingered in the science and engineering exhibits. His previous interest in machinery was strongly reinforced. The family, however, struggled to make ends meet, and when Christoph Barnickel, the husband of a cousin of Theodor's named Betty Barnickel, who lived in Augsburg, Germany, offered to take Rudolf in, his parents quickly agreed.

After his arrival in Augsburg, Rudolf was enrolled in a three-year program at the local technical school, where he happily immersed himself in chemistry, art, and—especially—a machine shop with a forge. He was in mechanical heaven. At the age of fourteen, Rudolf wrote to his parents to tell them that he had decided to become an engineer. After the Franco-Prussian War ended in early 1871, the rest of the Diesel family returned to Paris. Rudolf remained in Augsburg, where he subsequently graduated at the head of his class, and then journeyed to Paris to be with his family. But his stay there was cut short by the sudden death of his older sister, Louise, possibly from heart failure. Grief-stricken, the Diesels eventually agreed to let Rudolf return to Augsburg when Professor Barnickel and his wife, who had become quite fond of their young relative, again offered to take him in.

In October 1873, Rudolf enrolled in a mechanical engineering program in Augsburg, and as usual he ended up at the top of his class. It was in the physics laboratory at this school that Rudolf saw a mysterious device that almost certainly had a profound impact on his life. The device, a pneumatic lighter, was a small cylinder constructed much like a bicycle tire pump, but with a barrel made of glass, in which air was greatly compressed, causing a substantial rise in temperature—and a hot spark that could be viewed through the glass. The seeds of an idea had been planted. But it would take many years before those seeds germinated and grew into a practical invention.

The "Heat Engine"

Upon graduation, Rudolf was awarded a scholarship to attend the Technische Hochschule München (literally the Technical High School of Munich, but generally known as the Munich Institute of Technology), where he studied thermodynamics under Professor Carl von Linde (the inventor of the first reliable and efficient compressed-ammonia refrigerator). By the time Diesel arrived in Munich, his interests had expanded to include the arts, linguistics, and social theories of the day. It was in Munich that Diesel first began to seriously consider the possibility of developing what he called a "heat engine," one that would not need a spark plug to create combustion (the general idea perhaps inspired by the pneumatic lighter he had seen in Augsburg, and unquestionably by some of Professor von Linde's lectures). At the time, the steam engine was the main source of industrial power, but the engine's poor heat-utilization efficiency—which ran around 10 percent or less—bothered Diesel, who abhorred even the slightest waste. Steam engines were also extremely expensive and consequently tended to benefit large and wealthy companies that could afford them over small businesses that could not. This deeply troubled Diesel's newfound social conscience. From this time on, Diesel combined his interests in engineering and thermodynamics with his desire to help small business owners, which became one of the motivations for his subsequent work in developing his own engine.

After graduating from the Munich Institute of Technology in 1880, Diesel became an apprentice at the large and successful Gebrüder Sulzer Maschinenfabrik (Sulzer Brothers Machine Works) in Winterthur,

Switzerland, where he gained a good deal of practical mechanical experience. Diesel immediately impressed his supervisors, and it wasn't long before he was sent to Paris, where Carl von Linde (who was also a businessman) helped arrange a job for him in a new refrigerator manufacturing plant that was under construction. Diesel quickly became a refrigeration expert and was soon promoted to plant manager. In 1881 Diesel obtained his first patent (for the production of ice in glass containers), and a short time later he met his future wife, Martha Flasche, an attractive, blue-eyed German blonde who was a governess for a wealthy Parisian family. With his patent in hand, Diesel searched for a manufacturer for his ice-making machine. He soon came to an agreement with the Maschinenfabrik Augsburg (Augsburg Machine Works, known today as Maschinenfabrik Augsburg-Nuerenberg, or MAN AG), which began producing parts for his invention in 1883, the same year he married Martha in Munich. The next year, their first child, Rudolf Jr., was born, followed by a daughter, Heddy, and a few years later another son, Eugen.

In 1885 Diesel began to experiment with an ammonia-fueled engine in his own laboratory. Although the experiments were ultimately a failure, these early trials set the stage for his later engine development work. In 1890 Carl von Linde helped Diesel obtain a new franchise to distribute and sell Linde's refrigerators in northern Germany. As a consequence, the young Diesel family moved to Berlin. Not long after arriving in Berlin, Diesel came up with a concept for a new engine design, which was finally patented on February 28, 1892. The following year, he published his now-famous paper "Theory and Construction of a Rational Heat Engine to Replace the Steam Engine and Contemporary Combustion Engine." The paper described his invention as a "compression ignition engine" that could burn virtually any fuel, ignited not by a spark but by the extremely high temperature caused by highly compressing the air before the injection of fuel into the cylinder. It was a radical idea. Now the challenge was to turn his idea into an engine that actually worked.

A Prototype

Diesel searched for someone to help him build a prototype. He again turned to the Augsburg Machine Works, which agreed to assist in exchange for future engine sales rights for most of Germany. He also received the backing of the Friedrich Krupp Werke at Essen (the German steel producer better known for locomotive and arms manufacture in later years) in exchange for all German sales rights not already given to the Augsburg company. In addition, Diesel received financial backing from the Sulzer Brothers company in Switzerland, in exchange for an option on Swiss patent rights. The first prototype, constructed in 1893, consisted of a single 10-foot black iron cylinder with a large flywheel at its base. The engine, which was initially fueled by kerosene and later by gasoline, ran briefly, but it had to be shut down when the pressure gauge exploded.[2] Diesel and one of his assistants narrowly escaped injury. Although the engine had run, it had not generated enough power to sustain its own operation. Diesel went back to the drawing board and extensively redesigned the engine. He also continued to experiment with various fuels, and a second prototype actually ran briefly under its own power on February 17, 1894. But numerous technical problems still needed to be resolved. During this period, Diesel, who had always been prone to headaches, began to suffer severe migraine symptoms.

"When I began constructing my engine in the early nineties, the existing method was a total failure," Diesel later admitted. "The enormous pressures generated in my machine, the friction between moving parts, the magnitude of which had not been seen before, forced me to minutely examine the stress on each single organ and to delve extensively into material science."[3] This examination process delayed Diesel's project for over a year. In addition to making major redesigns of numerous components, he also continued his fuel experiments with gasoline, kerosene, lighting gas, and heavy oils.[4] Low-grade kerosene was the final fuel of choice. But the time he took to fine-tune his engine design was well spent. On the last day of 1896, a third prototype engine, containing many refinements over Diesel's earlier models, was started up and thoroughly tested.

The six-hundred-pound, fifty-seven-inch-high engine ran smoothly and was considered a success.

As word of the invention spread, Diesel began to sell licenses to entrepreneurs who wanted to manufacture his engine outside of Germany. Most of these manufacturers added technical refinements of their own to Diesel's basic design, and they were supposed to share their improvements with the other licensed manufacturers. But in practice, this often did not occur due to competitive pressures.

One of the most prominent of these businessmen was Adolphus Busch, the U.S. beer baron of Saint Louis, Missouri. After hearing about Diesel's new engine, Busch sent a trusted friend and technical advisor, Colonel Edward D. Meier, to Augsburg to study the engine and report his findings. Meanwhile, Busch settled into a luxury hotel in Baden-Baden, where he awaited the outcome of the investigation. Meier's eventual report was extremely favorable, and Rudolf Diesel was invited to Baden-Baden. Diesel quickly sensed that the high-spending Busch wanted the manufacturing license for his engine in the United States badly and wouldn't be stingy about paying for it. Diesel boldly suggested a fee of one million German marks (roughly equivalent to US$5 million today). Busch didn't bat an eyelash and calmly wrote out a check on the spot. Busch subsequently had the first commercial engine built in the United States on the Diesel patent installed in his Saint Louis brewery.

The Munich Exhibition

In 1898 four of Diesel's engines—each one built by a different German manufacturer—were put on display at the Munich Power and Machinery Exhibition. The special pavilion housing the exhibit was one of the highlights of the fair. Although the engines, which were constantly attended by technicians, ran without serious difficulties most of the time, there were bad days. One of the worst occurred when Prince Ludwig (the future King Ludwig III) of Bavaria visited the exhibit. The prince had heard that Diesel's engines were considerably less noisy than other engines of the day and wanted to see (and hear) them for himself. Unfortunately, none of the engines were working that day. The prince entered the pavilion and inspected the silent, inoperative engines. After a lengthy pause, he commented that the new engines were *indeed* quiet.

As more and more companies lined up to buy licenses, Rudolf Diesel quickly became a wealthy man. To help manage the exponential growth of the business, Diesel established the Allgemeine Gesellschaft für Dieselmotoren (General Society for Diesel Engines) on September 17, 1898. The new company paid him 3.5 million marks in cash and stocks for the rights to his engine and assumed control of all future development work. The following month, still suffering from severe headaches and exhaustion from his years of constant work, Diesel was urged by his doctors to enter a private sanitarium in Munich for a rest cure. But he was still too close to his business, and after the unsuccessful stay at the sanitarium, his physicians recommended a further course of treatment in a remote castle in the Alps. Although the second rest cure was a pleasant diversion in spectacularly beautiful surroundings, it was not entirely successful either.

It was during this turbulent period in Diesel's life that he had a sumptuous mansion built for his family. Located at Maria-Theresia-Strasse 32 in a fashionable suburb of Munich, this extravagant indulgence featured decorated vaulted ceilings, the most modern plumbing and electrical wiring, a marble fireplace in every room, and the finest French and Italian furnishings. It cost a fortune. Under normal circumstances the cost would not have been a problem for a multimillionaire, but Diesel, who had been so successful with his own business, seemed to have an uncanny talent for losing money in other people's ventures. He managed to lose 300,000 marks in a questionable Balkan oil scheme that went bad. He also was frequently engaged in expensive legal proceedings to protect his patents or defend his business interests. One legal judgment (involving some of his real estate speculations) cost Diesel 600,000 marks.

The Paris Exposition

In 1900 a smaller version of Diesel's engine manufactured by the French Otto Company was shown at the Paris Exposition. To demonstrate the engine's ability to operate on various fuels, it ran on peanut oil. Around fifty thousand people attended the exposition, and some of them were undoubtedly confused when they followed the scent of what they thought was a food concession and ended up standing in front of Diesel's engine chugging away on the peanut oil. Some people cite this exhibit as evidence of Diesel's early support of renewable biofuels. But the idea for the use of peanut oil appears to have come from the French government, which had its peanut-growing African colonies in mind. Diesel did conduct similar vegetable oil tests in later years, however.[5] In any case, the exhibit was extremely popular, and the engine won the exhibition's coveted Grand Prize. If nothing else, the prize was good publicity. By the end of 1901, some thirty-one companies were licensed to build and sell diesel engines and sales franchises covered eleven countries. A survey conducted the following year counted 359 diesel engines in use worldwide.

Unfortunately, Diesel's vision of an engine that would empower small businesses to compete with large industrial companies didn't come to pass in his lifetime. His early models required a separate air compression unit to inject the fuel, and the engines themselves were too large and expensive for all but the heaviest of industrial uses. If anything, Diesel's engine simply increased the competitive advantage of the big companies. However, oceangoing ships had plenty of space in their engine rooms, and Diesel's engine design was adopted for marine use as early as 1903. Despite these early successes, Diesel's persistent bad luck with investments continued to multiply. By 1905, his estimated losses came to over 3 million marks.

Ironically, Diesel came to hate his extravagant mansion in Munich, which he began to describe as an "outrageous mausoleum." But he did make it somewhat more practical by converting the entire upper floor into an engine research and construction office. Martha Diesel was not thrilled. Nevertheless, it was in his home office that Diesel planned a smaller version of his engine that would be suitable for automotive use. The "petite model," as he called it, earned a Grand Prize at the Brussels World's Fair in 1910. But this engine still had many shortcomings and was not immediately adopted for its intended use.

Vegetable-Oil Vision

In his later years Diesel was unquestionably a strong advocate for the use of renewable fuels such as seed oils, and this almost certainly was an outgrowth of his original desire to create an engine that could be operated almost anywhere on almost any fuel. In 1911 he said, "The diesel engine can be fed with vegetable oils and would help considerably in the development of agriculture of the countries which use it." The following year, despite the growing dominance of petroleum-based fuels, Diesel continued to make the case for vegetable-oil fuels. In an April 13, 1912, speech in Saint Louis, Missouri, Diesel described the recent changes to the fuel atomizers in his engines that allowed the use of castor oil, palm oil, lard, or other natural fuels. He went on to say, "The use of vegetable oils for engine fuels may seem insignificant today, but such oils may become, in the course of time, as important as petroleum and the coal-tar products of the present time. . . . Motive power can still be produced from the heat of the sun, always available, even when the natural stores of solid and liquid fuels are completely exhausted." Prophetic words indeed.

In November 1912, Diesel published a book, *Die Entstehung des Dieselmotors*, describing the development of his engine. The income from the book helped offset a little his mounting debts from continuing bad investments and various lawsuits, which by this time totaled nearly 10 million marks. Nevertheless, in early 1913 Rudolf and Martha vacationed in Italy. After their return, the summer of 1913 was not distinguished by any dramatic events, except that Rudolf seemed to be sinking into a deepening depression over his financial difficulties. On September 26, Diesel took a slow train to Belgium. He spent several days at the elegant Hotel de la Poste in Ghent, where he wrote a confused letter to his wife. Unfortunately, he misaddressed it, and she did not receive it until about five days after he disappeared. Then, on the evening of September 29, he boarded the steamship *Dresden* at Antwerp for his ill-fated voyage to England.

What really happened in the middle of the English Channel on that dark night in September 1913? No one will ever know for sure. Perhaps Diesel, distraught over his impending financial calamity, decided to take his own life. Diesel had been suffering from heart trouble for a number of years, and it's possible that he was struck by a sudden heart attack and fell overboard. Or, perhaps, the dark speculations about embittered enemies or international intrigue were correct. . . . Whatever happened, with Diesel's passing the world lost one of the most brilliant engineering minds of the day—and the father of what has become the biodiesel industry as well.

2

Vegetable Oil Revival

In 1912, a year before Rudolf Diesel's death, there were more than seventy thousand diesel engines operating around the world. But because of their size and weight and the fact that they ran best at a relatively constant speed, about 95 percent of these engines were restricted mainly to stationary uses in factories and electrical-generation plants.[1] Nevertheless, the use of diesels to power ships was beginning to gain some momentum. By 1912, around 365 ships and at least 60 heavy merchant vessels were equipped with diesel engines.[2] In the last five years of his life, Diesel had focused much of his research and development work on trying to build a successful diesel-powered locomotive—with little success. Nevertheless, he was convinced that his engine would revolutionize the railroad industry around the world. Diesel was right, but he didn't live to see it happen. The first diesel-powered locomotive ran on the Prussian and Saxon State Railways in 1914,[3] and early diesel locomotives in the United States finally made their debut around 1925. However, diesels did not seriously displace steam engines on most railroads until after World War II.

The automotive industry was particularly slow to adopt Diesel's engine. But eventually smaller, lighter engines with self-contained direct-injection pumps were developed, and in 1924 Germany's MAN AG was the first company to power a truck with one of these new engines. In the United States, the Mack Truck Company began experimenting with diesel engines around 1927, and by 1935 Mack was producing and selling their own diesel-powered trucks.[4] In the early 1930s, diesel trucks were soon followed by diesel-powered public buses in Berlin and many other cities. In the fall of 1936 Daimler-Benz unveiled their first large-scale-production diesel passenger car at the International Automobile Exhibition in Berlin. Although diesel-powered taxis subsequently became popular, particularly in Europe, due to their low fuel consumption and rugged dependability, the general driving public found diesel-powered pleasure cars of this era to be sluggish and unexciting. But for heavy commercial purposes, such as factories, power plants, oceangoing ships, railway locomotives, and construction equipment, diesel engines quickly became the motive power of choice—and remain so to this day.

Although Diesel had originally intended that his engine would be able to run on a variety of fuels, including whale oil, hemp oil, and coal dust, early on he opted for less expensive petroleum-based kerosene, which was plentiful and relatively cheap in the late nineteenth century. In the early twentieth century, as diesel engines came into wider use, the petroleum industry developed so-called diesel fuel, a lower-grade by-product of the gasoline refining process. Diesel engines were modified to burn this new fuel. But the more Diesel's engine was altered to burn this dirtier, petroleum-based fuel (petrodiesel), the further it diverged from its inventor's vision of biofuel flexibility, which he had especially promoted in the last few years of his life. And as the number of diesel engines that were manufactured to run on petrodiesel grew exponentially, the less likely it seemed that Diesel's vegetable-oil vision would ever be realized.

Early Experiments

Despite the development of an almost total reliance on petrodiesel, the idea of using vegetable oil as an alternative source of diesel fuel was not completely forgotten. There were a number of experiments on the use of vegetable oils beginning in the 1920s and continuing through the early 1940s, particularly among European nations with tropical colonies—especially African colonies. In 1937 a Belgian patent was granted to G.

Chavanne of the University of Brussels for the use of ethyl esters of palm oil (which would now be described as a type of biodiesel). The following year, a commercial passenger bus operated between Brussels and Louvain on palm oil ethyl ester. The test was reportedly a success.[5] Shortly before World War II there were some experiments in South Africa in which heavy farming equipment was operated with fuels derived from sunflower seed oil. But these promising experiments were abandoned in favor of making synthetic liquid fuel from coal, due to South Africa's abundant coal reserves.[6]

During World War II, vegetable oils were used as emergency fuels by various nations when normal supplies of petroleum-based fuels were disrupted. Brazil utilized cottonseed oil in place of imported diesel fuel. Argentina, China, India, and Japan all used some form of vegetable oil during the war years. The Japanese battleship *Yamato* is reported to have used refined, food-grade soybean oil as bunker fuel.[7] But after the war, with the return of steady supplies of cheap petroleum oil, virtually all research on vegetable-oil fuels ceased.

The Oil Embargo

Fast-forward to 1973. On October 6 of that year, Egyptian military forces attacked Israel across the Suez Canal, while in the north Syrian troops simultaneously attacked from the Golan Heights. Initially taken by surprise, the Israeli army pulled back, but then with assistance from the United States and other Western nations, the Israelis counterattacked and eventually reversed all the early Arab gains and ended up occupying additional territory. Although a cease-fire was concluded in November, the Arab nations were humiliated by their losses—and angry. In retaliation for the United States' and other Western countries' support for Israel, the Organization of Petroleum Exporting Countries (OPEC) initiated an oil embargo against the West in general, and the United States in particular.

The embargo sent shock waves through the global economy and dramatically inflated energy prices worldwide. Overnight, the price of a barrel of oil rose from $3 to more than $5. By the end of 1974 the price was over $12 a barrel. In the United States, President Richard Nixon instituted voluntary gasoline rationing in December 1973 and urged homeowners to turn down the thermostats on their heating systems. In response to the crisis, the U.S. Congress approved the construction of the trans-Alaskan oil pipeline (completed in 1977 at a cost of $8 billion), which was designed to supply 2 million barrels of oil a day. Much of the Western world was hit by a severe economic recession, but western Europe was especially hard hit because it had very limited domestic supplies of oil at the time.

In 1979 the revolution in Iran that resulted in the ouster of the U.S.-backed Shah precipitated yet another global energy crisis. Oil prices again doubled, sending the industrial world into an economic tailspin.

The price increases and fuel shortages of the 1970s and early 1980s spurred interest in the development of alternative fuels around the world, but especially in western Europe and the United States, where the economic damage from the turmoil in the petroleum markets was most severe. One of these alternative fuels was ethanol. During both World War I and World War II, the United States and European nations used alcohol fuels as supplements to petroleum-based fuels. In the 1920s and '30s, a considerable amount of ethanol, fermented from corn and generally referred to as "gasohol," had been produced in the United States for use as a vehicle fuel. But this effort ended with the low petroleum prices of the 1940s, and as ethanol production facilities closed, gasohol disappeared from the market. But in 1979, in response to the ongoing international oil crisis, ethanol-gasoline blends were reintroduced to the U.S. market by several oil companies, which promoted the ethanol blends as "gasoline extenders" and octane enhancers. And while ethanol was being promoted as a gasoline additive, a separate but related line of research into alternative diesel fuels made from vegetable oils was being conducted in a number of countries.

Some of the earliest documented experiments with vegetable oils as diesel fuels in the United States took place in the early 1950s at Ohio State University, where a "dual fuel" project was conducted with cottonseed oil and corn oil blended with petroleum diesel.[8] But with the resumption of seemingly endless supplies of petroleum fuels after World War II, these studies soon gathered dust. After the shocks to the oil industry of the 1970s, however, interest in alternative diesel fuels was revived, especially in Europe and the United States. Two long-term research programs stand out in particular: one in Austria, the other in Idaho. A third program in South Africa showed a lot of initial progress but was not sustained.

The Austrian Connection

About a year after the energy crisis of 1973, representatives from the Austrian Federal Institute of Agricultural Engineering (Bundesanstalt für Landtechnik, or BLT) began preliminary discussions on alternative biofuels for diesel engines (particularly for farm tractors) with the diesel engine developer AVL-List GmbH in Graz, Austria. Those discussions were the impetus for a series of experiments that mixed various vegetable oils with petrodiesel. In 1975 Manfred Wörgetter, who had just graduated from the Mechanical Engineering Department of the Technical University of Graz, took a job at BLT in Wieselburg, Austria, and a short time later began the fuel experiments. "In my work I concentrated on bench and field tests of farm diesel engines, while my boss, Josef Pernkopf, focused on production issues," Wörgetter (who is now deputy director and head of Research Agricultural Engineering at BLT) recalls. "The main aim of our work was to ensure the supply of fuel for farm tractor engines in the event of another oil crisis."[9]

In 1976 and again in 1978, a small, old tractor was bench- and field-tested with various mixtures of linseed oil and petrodiesel. The initial results from the mixing tests showed that the viscosity (thickness) of the vegetable oil needed to be reduced and that the use of the oil would considerably increase engine maintenance costs due to excessive deposits on a number of internal engine components. In virtually all cases, if the vegetable oils were used over a long period of time, the engine could be seriously damaged. In early 1979 BLT published an article in a regional journal recommending that engine tests with linseed oil be discontinued. "We concluded that either the fuel needed to be adapted to the needs of the engine, or the engine needed to be adapted to the fuel," Wörgetter says. "We decided to adapt the fuel."

At this point the BLT researchers reviewed plant-oil production worldwide and concluded that a vegetable-oil strategy needed to be oriented toward local climate and agricultural conditions. "We concentrated our efforts on bench tests with different mixtures of rapeseed and sunflower oil with fossil diesel," Wörgetter says. "After four hundred operating hours with a farm tractor using 50 percent rapeseed oil in diesel we had to stop because of an engine failure." During this time, a copy of a South African newsletter containing information about work that had been done with sunflower methyl ester (a form of biodiesel) was brought to Wörgetter's attention. When he met some of the members of the South African research team at an energy conference in Berlin, Germany, in October 1981, Wörgetter's discussions with them convinced him that producing methyl ester from vegetable oil was probably the best approach for adapting the fuel for a diesel engine. The report of the use of sunflower oil methyl esters at this conference is generally viewed as marking the rediscovery of what came to be known a few years later as biodiesel.[10] In 1981 BLT published an article in a highly regarded Austrian journal on the idea of chemical modification of vegetable-oil fuels. At about the same time, some early feasibility studies on the use of fatty acid methyl esters as fuel in diesel engines and heating boilers were conducted in France, but most observers credit the sustained Austrian research work as being the main foundation for the subsequent European biodiesel industry.

Independently, and at the same time, a number of chemists at the University of Graz were having discussions in their laboratory about the possibility of using vegetable oil as fuel. They wanted to start a research project on that topic and contacted the Austrian Ministry of Agriculture to see if it was interested. The ministry informed them that BLT in Wieselburg was already working on the same topic, so a decision was made to contact BLT. Professor Martin Mittelbach, an organic chemist from the university, accompanied by his boss, Professor Hans Junek, traveled to

Wieselburg. "When we arrived, they told us they were having a bad day because the engine in one of their test tractors had stopped running," Mittelbach recalls. "They said, 'Maybe you have come at the right time; you are chemists, tell us what's going on.'"[11] Mittelbach says that he offered some immediate observations about the probable causes of the engine difficulties. He also agreed to investigate the problem further and promised to let the BLT researchers know if he discovered anything interesting. This meeting marked the beginning of a long-lasting friendship between the researchers. Mittelbach returned to Graz and gave a good deal of thought to the best ways of modifying the vegetable oil chemically to make a fuel that would not cause operational problems in diesel engines.

Mittelbach, who had received his PhD in 1979 at the University of Graz, says that shortly before he visited BLT in Wieselburg, someone in his department had mentioned the idea of using vegetable oil as a diesel fuel. "If you are an organic chemist you know that vegetable oil and diesel fuel have a totally different chemical structure, so it was hard to believe that vegetable oil could be used as a fuel," he recalls. "For nonchemists all oils may sound like they are the same, but they really aren't, so this looked like an interesting idea." Mittelbach checked the available literature to see what was available and didn't find much. (Mittelbach heard about Rudolf Diesel's work with vegetable oils a few years later but admits that this did not influence his early experiments because he was unaware of it at the time.)

The Process

Mittelbach persisted in his research and then conducted a series of chemical laboratory experiments and tests using rapeseed oil (also known as canola oil), which turned out to be successful. In chemical terms, vegetable oil normally is composed of three fatty-acid molecules linked to a glycerol molecule; combined in this way, the fatty acids and glycerol are referred to as a triglyceride. Mittelbach relied on a standard chemical process known as transesterification, in which the vegetable oil, an alcohol, and a catalyst were mixed, resulting in the removal of glycerin from the vegetable oil, to make the oil thinner. The products of the reaction are alkyl esters (biodiesel) and glycerin. The main point of the process was to produce alkyl esters that would flow through and combust properly in modern diesel engines without leaving damaging internal-engine deposits. "The fatty acids and fatty acid esters were not new," Mittelbach notes. "They were well known in chemical literature for making detergents, and then non-ionic detergents, which are used for dishwashing and so on. What *was* new was their use as a diesel fuel."

Another goal of this early research was to change the production process, because the existing oleochemical route for the production of fatty acid methyl esters was too complicated and expensive (oleochemicals, which are used in a wide range of products, are derived from biological oils or fats). The researchers wanted to find a low-temperature and low-pressure process that would perform the reaction without all the expensive equipment used by the traditional oleochemical industry. The research work finally resulted in a patent application that described the first method for cheap biodiesel production.

After the initial success with the experiments, Mittelbach and Wörgetter agreed that they needed a larger supply of biodiesel in order to conduct additional tests on the fuel. "At the university we were experienced chemists, but we were not set up to produce large quantities in the laboratory," Mittelbach says. "We would make a small batch and then have to clean everything up and do it again and again. It just was not practical." After receiving a small grant for the necessary equipment, the researchers were able to produce biodiesel in 30- to 40-liter (8- to 10-gallon) batches. When they had accumulated several hundred liters, they were ready for more extensive laboratory and field tests. The researchers looked for a financial partner for the field testing. The Austrian tractor manufacturer Steyr stepped forward. Steyr agreed to run the field tests while BLT would conduct the engine bench tests. The subsequent tests were successful, with no engine damage detected, and Wörgetter and Mittelbach published their findings in 1982.

Although the tests using rapeseed oil had been a technical success, it was quickly realized that the oil was too expensive as a feedstock to be used for biodiesel, since petrodiesel prices were much lower. At the time, very little rapeseed was being grown in Europe and most rapeseed had to be imported from abroad. Consequently, in 1983 Mittelbach began to look for less expensive feedstocks, and he soon discovered that there was a fairly large supply of used cooking oil available. He conducted some preliminary experiments using the waste cooking oil and found it to be a viable source for making biodiesel.

Mittelbach says that the most exhilarating moment for him, personally, was when he decided to run his own car on biodiesel. "During the early tests we made the biodiesel and sent it out to be tested; I didn't actually see the tests," Mittelbach recalls. "The tests were working, and so I thought, 'Well, what about passenger cars?' At the time I had a diesel car, and the most exciting moment for me was when I put around 20 liters of biodiesel in the fuel tank. I started the engine and checked the exhaust gases, which smelled like burned fat. It was very exciting. At that time, nobody could give a guarantee that your car would run on biodiesel. So, the next day, when it started and ran normally, I could see that it was going to work. We could have found a test car somewhere, I suppose, but I just decided to use my own car. That was the best proof that it worked. It was easier to convince other people that it worked because I could say that I was using it in my own car without any problems."

Real-World Tests

Now that the viability of biodiesel for use in a diesel engine had been demonstrated, the researchers decided that it was time to expand the tests to larger numbers of tractors in real-world settings. The idea was to persuade local farmers to use biodiesel in their tractors, creating a closed energy loop in the agricultural community. The farmers could grow and use their own fuel and be independent from the fluctuations in the international oil market. But in order to do that, much larger quantities of biodiesel needed to be produced. The solution was a pilot plant.

But the researchers didn't have the money for a venture of this size and were forced to look for an outside source of funds. The Wieselburg and Graz groups began searching for funding. Finally Mittelbach was successful, finding support in an unlikely place: the petroleum industry. Ironically, the main financial backing came from OMV AG, Austria's largest oil and petrochemical industry, according to Mittelbach. At the time, OMV was building a new oil pipeline from the Adriatic Sea to Vienna, and the money they paid for the acquisition of the land traversed by the pipeline was put into a fund to be used for the benefit of the farmers whose land was taken for the right-of-way. "The money was to be used for some sort of energy projects, and the farmers decided that the biodiesel pilot plant was one of the projects where the money should be spent," Mittelbach says. The pilot plant (which no longer exists) was constructed in 1985 at the Silberberg Agricultural College in Styria, Austria, and it was capable of producing around 500 tons (142,500 gallons) of biodiesel from rapeseed oil annually.

Once the pilot plant was up and running, the researchers discovered they had another problem. The farmers who were supposed to run the field tests had suddenly become skeptical. "We wanted to conduct fleet tests, and trying to find ten or twenty farmers who were willing to make the tests was not easy," Mittelbach says. "They were very anxious that something might happen to their tractors." But eventually, with a lot of gentle persuasion, the farmers agreed to the tests, which, happily, turned out to be successful.

In 1986 BLT was contacted by an official from Gaskoks, a large Austrian energy company, who wanted to help finance a pilot project, titled "Biodiesel," designed to create a basis for supplying the agricultural sector with fuel produced from rapeseed oil. The project, which was subsequently carried out at BLT in Wieselburg, was also financed and assisted by the Austrian Federal Ministry for Agriculture. During the project (which cost in excess of US$1 million), fleet tests were conducted on thirty-five farm tractors from ten different producers between 1987 and 1989. A series of bench tests, extensive emissions measurements, and investigations of the engine oil, along with the results of other Austrian research programs, finally led to a proposal for a biodiesel quality standard.

Thanks to the cooperative efforts of all the biodiesel researchers, the Austrian Standardization Institute (Osterreichische Normungsinstitut, or ON) established a working group that succeeded in creating the first biodiesel standard in the world (ON C 1190). "All subsequent standards have been based on the foundation set by this groundbreaking work," says Wörgetter.[12]

Turning Waste into Fuel

In 1987, building on the previous successes with rapeseed oil and Mittelbach's earlier experiments with used cooking oil, a series of engine and emissions tests using methyl ester from waste cooking oils was conducted in cooperation with AVL-List GmbH in Graz. "We used the same chemical reactions that we had used with the rapeseed oil," says Mittelbach. "We also measured the emissions, because if you are using an alternative fuel this is an important issue. The soot and particulate emissions with the vegetable oil biodiesel were far lower—about 50 percent lower—than with the petroleum diesel. And the used frying oil was even better than rapeseed oil biodiesel. That was really surprising; we had not anticipated that. This was a very important step for the further development of biodiesel." In 1988 the results of the tests were published in *Science News* and the *Journal of the American Oil Chemists' Society*. The international scientific community began to pay a lot of attention. In the same year the first Austrian patent for the transesterification process was taken out by Mittelbach and several collaborators. Although the research teams in Graz and Wieselburg worked well together in the early years, eventually a certain amount of competition developed between them, especially from around 1987 to 1990, according to Wörgetter. But since then, a good, cooperative relationship has been reestablished, according to members of both teams.

In 1989 an Austrian-government-supported research project, "High-Quality Fuel from Waste Cooking Oils," was launched. A short time later, large-scale esterification experiments followed using 100 percent waste cooking oil, conducted by Mittelbach and others, in cooperation with the Technical University of Graz. But not all the emphasis was on used cooking oil. In 1990 the first farmers' cooperative in Asperhofen (near Vienna), with approximately 290 farmer members, began commercial production of biodiesel made from rapeseed as well as sunflower oil. In the same year, large-scale tractor engine tests were finally completed by

BLT at Wieselburg, which convinced major tractor manufacturers such as John Deere, Ford, Massey-Ferguson, Mercedes, Same, and others to issue engine warranties for biodiesel use.[13] The warranties were a major step forward in the development of a viable European biodiesel market.

Although a group of Austrian farmers were participants in some of the earliest large-scale field trials of biodiesel, the farming sector was generally slow to adopt biodiesel for regular use. "The idea sounded very attractive and was well accepted by politicians, scientists, the general public, and the tractor industry, but less well accepted by the farmers who were just looking at the cost," says Werner Körbitz, the chairman of the Austrian Biofuels Institute (Osterreichisches Biotreibstoff Institut, or OBI) in Vienna. "If it was one penny more expensive, they were not interested. If it was one penny less expensive, then they were willing to try it. There was very little market research done about customer behavior and attitudes at this early stage. It was thought that biodiesel would be produced by farmers for farmers, but in actual practice, this was not true."[14] In later years, however, even the reluctant agricultural community began to embrace biodiesel.

South Africa

The story of biodiesel research in South Africa is a tale of missed opportunity—on more than one occasion. As mentioned previously, there were some early experiments in South Africa prior to World War II using vegetable-derived fuels, which were abandoned in favor of synthetic liquid fuels produced from coal. Then, around 1980, the idea of using vegetable oil as a diesel fuel was revived at the Council for Scientific and Industrial Research (CSIR) in Pretoria, South Africa. Lourens du Plessis, now a semiretired special research scientist in the food science group at CSIR, recalls what happened.

"One of our engineers did some early vegetable-oil tests on a diesel engine on our campus," he says. "The agricultural engineers quickly realized that this was a good idea that could benefit the agricultural community."[15] The CSIR research project began with investigations into the use of straight sunflower oil as a fuel. But the experiments soon ran into problems with improper fuel vaporization and the leaking of fuel into the lubricating oil inside the engine. "So, we thought about how we could overcome these problems, and we quickly switched over to producing ethyl esters of sunflower oil," du Plessis continues. "I was in favor of methyl esters, but the engineers wanted to stick to ethyl esters because they regarded it as an agricultural product. The whole issue at the time was for the agricultural sector to completely produce the diesel fuel." Du Plessis, who had degrees in chemistry and botany as well as a PhD in plant biochemistry, was asked to be part of the research team because of his expertise in oils and fats.

The initial test results were positive, and du Plessis and his coworkers produced about 500 liters (132 gallons) of the biodiesel fuel in a processor housed in a small steel shed. "It wasn't a technically designed facility," he recalls. "We built it ourselves in the backyard; it was actually quite a lot of fun to do it all ourselves." With a relatively large supply of fuel on hand, the agricultural engineers ran the engine tests, mostly on diesel tractor engines, in their laboratories. The results of the tests were encouraging, with very few problems of any sort. An additional series of tests was run on the stability of the biodiesel fuels. The results of all the tests were published in various journals and presented at a number of international conferences. Then, around 1984 or '85, the South African Department of Agriculture pulled the plug on the program, according to du Plessis. "They decided it was not economic or worthwhile, and they stopped the whole process," he says. "And that was the end of it."

Du Plessis looks back on his early biodiesel research with mixed emotions. "What was really rewarding was the fact that as a food chemist and scientist, I worked with the mechanical engineering people; it was a multidisciplined team," he says. "But the main thing that was so exciting for us was that the fuel was running in the engines without any problems. We actually tried to patent our ideas in the early 1980s, but we found that there was some early Belgian work already registered, so we couldn't get a patent. That was a pity, because we spent a lot of time and made a lot of effort to get the fuel well tested." But the saddest part of all was that the research program, which had made such progress and showed such promise, was canceled altogether.

Early Experiments in Idaho

The oil shocks of the 1970s sparked considerable interest in alternative fuels on the other side of the Atlantic Ocean as well. The initial research work with vegetable-oil diesel fuels in the United States began at about the same time as the Austrian and South African studies were getting underway. There were a number of early studies in the late 1970s, but one of the main pioneers in this work was Dr. Charles Peterson, an agricultural engineer at the University of Idaho. In 1979, the dean of the College of Agriculture at UI told Peterson he had heard that vegetable oil could be used as a diesel fuel. Peterson was intrigued, and he agreed to give it a try. He went to a grocery store and bought some sunflower oil, which he mixed with regular petroleum-based diesel fuel. "We had a Ford tractor that we used in class experiments, so it had an easy way to switch fuels," Peterson, who is now emeritus professor of biological and agricultural engineering at UI, recalls. "We had a dynamometer that we put it on, but I think the biggest thing we learned was that the vegetable oil blended very well with the diesel fuel."[16]

The following summer, encouraged by the initial tests, Peterson decided to try running the tractor on straight safflower oil in demonstrations at a number of local county fairs. "We used safflower oil because in testing the viscosity of the oils that we grew in Idaho, safflower oil had the lowest viscosity, and we thought it might perform a little better," Peterson says. That experiment worked too—at least initially. "When we got done, however, the tractor engine was completely shot," he admits. "At the end of the test we had severe polymerization of the piston rings and the engine wouldn't start."

Peterson's experiment with straight vegetable oil didn't work any better than other similar experiments of that period because modern diesel engines simply weren't designed to run on vegetable oil. Consequently, in order to use vegetable oil successfully as a fuel, either the oil or the engine had to be modified. Peterson and his colleagues, like their counterparts in Austria and South Africa, quickly came to the conclusion that modifying the oil was the preferred approach. Trying to develop the best ways to do that, and then testing the performance of the resulting fuels, became the main focus of UI's subsequent, highly regarded biodiesel research program.

Around 1982, the university bought a new Sato tractor for its farm operation (it is still in use today). "We started out using a 50 percent rapeseed oil/fuel blend for the tractor, but shortly afterward we switched it to run on 100 percent esterified rapeseed oil," Peterson recalls. The switch to esterified oil (biodiesel) marked the real beginning of biodiesel experiments at the university. The UI Chemical Engineering Department helped Peterson come up with his early biodiesel recipe. But even the best-designed plan can go astray. "I remember when I tried to make the very first batch of esterified oil, I miscalculated the catalyst and ended up with a big glob of soap," Peterson recalls, laughing. "People still do that sometimes, I guess." Undaunted, Peterson persisted, and soon he and his coworkers were regularly turning out high-quality biodiesel from rapeseed oil. The process was facilitated with the help of a small mechanical screw press for extracting the seed oil and a 200-gallon batch reactor to make the biodiesel. Since then, the university has produced thousands of gallons of biodiesel for its many research projects.

Another early "disaster" from the mid-1980s that eventually led to a research breakthrough was an engine test with raw vegetable oil conducted by one of Peterson's graduate students. "He did a power test with raw vegetable oil in an engine connected to an electric dynamometer," Peterson recalls. "In a very short time it polymerized the piston rings and the engine seized up. This gave us the idea of running what we called the 'injector coking test' as a screening method for evaluating alternative fuels. We would run the engine through a torque test, and at the end of that period we pulled the injectors and evaluated the coking [carbon deposits] on them. The coking on the injectors could then be related to the coking that was going on inside the rest of the engine. This was a screening test that was a lot easier on the engines. We felt that the fuels that performed better in the test would do better on longer-term tests as well. There have been a lot of people who have adapted the test in different ways, and now they use machine vision to do the evaluation. When we did the tests we used photographs to determine the extent of the coking. It's interesting that this test came out of what was a failure and pretty severe damage to that engine."[17]

The Research Expands

Before long, the rapeseed biodiesel project became a joint effort with the Idaho Department of Water Resources, the U.S. Department of Energy, and the U.S. Department of Agriculture. Over the years, Peterson, who is widely acknowledged as a leader in the field, has headed numerous biodiesel research projects for various local, state, federal, and private agencies. During that time, UI has conducted biodiesel research on rapeseed oil, soybean oil, hydrogenated soybean oil, tallow, and a number of other feedstock sources.

The University of Idaho has used different forms of biodiesel to run many diesel engines in various types of farm machinery, stationary installations, Dodge and Ford pickups, and a long-haul Kenworth truck. The 200,000-mile, over-the-road test on the Kenworth tractor-trailer, operated by Simplot Transportation of Caldwell, Idaho, was completed in 1999. The truck ran on a blend of 50 percent biodiesel and 50 percent petrodiesel. The biodiesel was made from waste vegetable oil from the Simplot Inc. french-fry plant located in Caldwell. After the test, the Caterpillar engine was removed from the truck and sent to the manufacturer for evaluation. The entire test was considered to be a success. "I think that test, which involved Caterpillar, really helped get their attention focused on biodiesel, and it probably was as important as anything in making people aware that waste oil could be used for biodiesel," Peterson says.[18]

Another research initiative that Peterson is especially proud of is the "Truck in the Park Project" conducted in Yellowstone National Park. The project had two main goals: to provide data on emissions and performance and to define a niche market for biodiesel in an environmentally sensitive area. The project also developed partnerships among the U.S. Department of Energy, the states of Montana and Wyoming, the National Park Service, regional businesses, and regulators. In 1994, Peterson and Darly Reece, a graduate student working on his master's degree in agricultural engineering, drove a new Dodge pickup fueled with 100 percent rapeseed ethyl ester (REE) to Mammoth Hot Springs. Due to the park's extremely cold winter-weather conditions, the truck was equipped with a standard winterization package for diesel engines, but no modifications were made to the truck's engine or fuel system.[19] Initially there were some concerns that the odor of the biodiesel exhaust (which smelled like french fries) might attract bears. A special bear study was conducted, and it was found that the bears couldn't have cared less. The truck was then driven by park employees, who accumulated more than 130,000 miles running on the biodiesel fuel. In September 1998, the truck's engine was completely torn down and thoroughly inspected, revealing very little wear and no carbon buildup. The engine was reassembled and put back in the truck for the second phase of the test, designed to accumulate 200,000 miles of use. Overall, the project has been a great success with park employees and visitors alike, and the National Park Service has since introduced biodiesel to twenty other parks through the Green Energy Parks Program.

"That certainly was a dramatic setting for the tests," Peterson notes. "I think our Yellowstone Park project, where we used biodiesel in such an environmentally sensitive area, is probably the project that made biodiesel as well known in this country as it is." Looking back on his quarter century of work with biodiesel, Peterson is somewhat bemused by the present scope of the program. "I always tell people that this started out more as a hobby than anything," he says. "In the early years, I never ever thought that this would develop into our principal research program that I was involved with. But after a while, we were able to get more funding for this project than for others, and it kept on getting larger and larger." Although it didn't occur until about ninety years after his death, Rudolf Diesel's vegetable-oil vision was finally being realized.

3

Biodiesel 101

As most of the early biodiesel researchers quickly discovered, using straight plant oil as a fuel substitute in diesel engines was not especially good for the engines. Numerous test engines around the world were undoubtedly ruined in many of those early experiments. The problem was that, for almost a century, diesel engines had been gradually developed, adapted, and fine-tuned to burn petroleum-based diesel fuel. Faced with the prospect of having to modify the millions of diesel engines in use around the world in a wide range of different types of vehicles, researchers in the late 1970s and early 1980s opted to modify the vegetable-oil fuel instead. This was actually a fairly easy choice, according to Werner Körbitz, the chairman of the Austrian Biofuels Institute in Vienna. "That's because, even then, it was clear that biodiesel would not be able to replace more than about 10 percent of the petroleum diesel market share," he says.[1] And trying to retool the diesel engine for such a small potential part of the market didn't generate much enthusiasm among engine manufacturers.

Making Biodiesel

As mentioned previously, the oil transformation process the researchers selected was transesterification, or the transformation of one form of ester into another. (Esters are naturally occurring compounds such as oils and fats, or any of a large group of organic compounds formed when an acid and alcohol are mixed.) In order to understand the process (without getting too technical), we need to take a closer look at vegetable oil. One of the main problems with vegetable oil compared to diesel fuel is that it's thicker, or more viscous. This is due to the fact that vegetable oil contains glycerin—a thick, sticky substance—in its chemical structure. Every vegetable-oil molecule is composed of three fatty acid chains attached to a molecule of glycerin. Picture a microscopic three-legged creature with a round glycerin head and three long, dangling legs. This is why vegetable oil is described technically as a *tri*glyceride, or *three* fatty acid chains and glycerin.

Although the exact percentage varies somewhat depending on what kind of plant the oil comes from, approximately 20 percent or less of a vegetable-oil molecule is composed of glycerin. Transesterification involves breaking every oil (triglyceride) molecule into three fatty acid chains and a separate (or free) glycerin molecule. During the process, alcohol is added, and each of the fatty acid chains attaches to one of the new alcohol molecules, creating three mono-alkyl esters. This process makes the esters thinner and more suitable for use as diesel fuel. Once separated from the glycerin, the alkyl ester chains are what is called biodiesel.

The alcohol used in the process can be either ethanol (made from grains) or methanol (made from wood, coal, or natural gas). Methanol is usually preferred because it's cheaper and tends to produce a more predictable reaction. On the downside, methanol dissolves rubber, can be fatal if swallowed, and must be handled with extreme caution. Ethanol, on the other hand, is generally more expensive and may not always produce a consistent, stable reaction. On the upside, ethanol is less toxic and is made from a renewable resource. If biodiesel is produced with methanol it is referred to as *methyl esters*, and if it is made with ethanol it is referred to as *ethyl esters*. A more generic term, *alkyl esters*, refers to any alcohol-produced vegetable-oil esters.

But as a good high-school chemistry teacher would point out, there is still one more ingredient needed to make the process work—a catalyst. The catalyst is the substance that initiates the reaction between the vegetable oil and the alcohol by "cracking" the triglycerides (vegetable oil) and releasing the alkyl esters (biodiesel). There are two main catalysts that can be used, sodium hydroxide (NaOH) and potassium hydroxide (KOH). Sodium hydroxide, which is less expensive, is commonly referred to as lye or caustic soda (the same chemical used to unclog kitchen or bathroom drains). If sodium hydroxide is not available, potassium hydroxide can be used instead, but a larger quantity is required. Sometimes a third catalyst, sulfuric acid, is used by commercial biodiesel producers as a pretreatment for waste cooking oils to prevent excessive soap formation. All of these chemicals are dangerous, however, and must be handled carefully.

The Process

The transesterification process is initiated by adding carefully measured amounts of alcohol mixed with the catalyst to the vegetable oil. How much catalyst is used depends on the pH (acidity) of the oil. When used cooking oil is chosen as the feedstock for biodiesel, one additional factor comes into play—free fatty acids. Because they are considered acids, oils and fats are sometimes referred to as fatty acids. When vegetable oil is fried, free fatty acids (which are not bound or attached to other molecules) are released and end up floating around among the triglycerides. These free fatty acids can use up too much catalyst and result in the formation of excess amounts of soap (an unhelpful trait), so they need to be eliminated. The way to accomplish this is to add more catalyst to the mix; the exact amount is determined by the pH of the used cooking oil or by the trial and error method.

One of the main advantages of biodiesel is that the transesterification process used to produce it can be conducted at almost any scale—from a kitchen blender that makes a few liters on up to a large industrial facility capable of producing millions of gallons per year. Although an industrial-size biodiesel facility uses a lot of high-tech equipment to wring every last productive ounce out of all the ingredients (and recycles many of them for reuse), the basic transesterification process is more or less the same as that used in a small-scale facility located in a garage or backyard shed. The main difference is that very large-scale operations are often designed to produce biodiesel on a continuous basis—the *continuous-flow process*—

while the small processor normally produces smaller, individual batches at a time—the *batch process*. In the batch process the reaction and subsequent settling procedure takes place in a single tank or container over a period of time. In the continuous-flow process, however, there is a constant movement of feedstock and other ingredients through the system, resulting in finished biodiesel at the end of the process.

Here's how the basic process works (using methanol and sodium hydroxide as an example). Carefully measured quantities of methanol and sodium hydroxide (lye) are mixed to create sodium methoxide, which is then mixed with the vegetable oil and stirred or agitated (and sometimes heated) for a specified length of time. If used vegetable oil is the feedstock, the process requires a bit more testing, lye, and filtration, but it is otherwise essentially the same. During the mixing, the oil molecules are split or "cracked" and the methyl esters (biodiesel) rise to the top of the settling/mixing tank while the glycerin and catalyst settle to the bottom. (The separation process can be speeded up with the use of a centrifuge.) After about eight hours, the glycerin and catalyst are drawn off the bottom, leaving biodiesel in the tank. In most cases the biodiesel needs to be washed with water to remove any remaining traces of alcohol, catalyst, and glycerin. In this procedure, water is mixed with the biodiesel, allowed to settle out for several days, and then removed. The wash process can be repeated if needed, but it is time-consuming. Not everyone agrees on whether the water wash is necessary. Some smaller producers who are making biodiesel for themselves skip the process, while commercial producers usually must do so to meet industry standards. In the case of some larger, more sophisticated manufacturing facilities, the transesterification process itself is so carefully controlled and refined that the water wash may not be needed. There are, of course, quite a few technical variations on this entire process for large-scale industrial operations, but the general transesterification procedure is similar.

Because making biodiesel is relatively simple and can be very low-tech (an old 55-gallon drum is often used as the settling/mixing tank), it has attracted an enthusiastic community of backyard enthusiasts or "home brewers" around the world. For those who want to make their own biodiesel, *From the Fryer to the Fuel Tank: The Complete Guide to Using Vegetable Oil as an Alternative Fuel* by Joshua Tickell is essential reading. A newer reference that goes into more detail is available from Maria "Mark" Alovert, called the *Biodiesel Homebrew Guide 2004* (see the bibliography).

Biodiesel Feedstocks

Another remarkable feature of the transesterification process is that it can use a wide range of feedstocks—virgin vegetable oils, used fryer oil, animal fats, even pond algae—to produce the same basic biodiesel end product (with minor differences in fuel characteristics). These feedstocks can be used individually or blended to produce biodiesel with specific traits. The ability to adapt the production process to locally available feedstocks and end-user needs is one of biodiesel's most attractive advantages.

Oil-Producing Crops

There are hundreds of oil-producing plants that can be used as feedstocks for biodiesel, from corn to soybeans and sunflower to oil palm. Even avocado and industrial hemp will work. Here is a brief overview of some of the most common crops, listed in descending order of oil production in liters per hectare (2.47 acres) and in U.S. gallons per acre.[2] The production figures given are conservative averages and can vary widely depending on specific plant variety, cultivation practices, and weather and soil conditions. Other products made from these plants and their oils are included in the listings to highlight existing competitive markets. Figure 2 shows the main feedstock crops used worldwide for biodiesel production.

Oil Palm

The African oil palm is at the top of the list of oil-producing plants, with a remarkable yield of up to 5 metric tons of oil per hectare every year (about 5,950 liters per hectare, or 635 gallons per acre). The African palm produces two types of oil, palm oil and palm kernel oil. Palm oil is

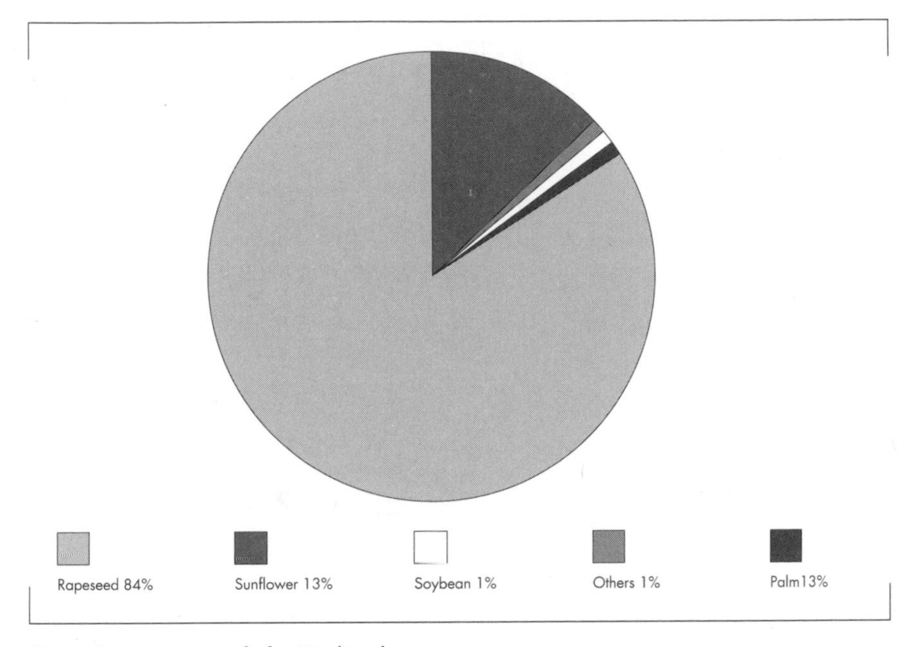

Figure 2. Raw Materials for Biodiesel (from the Austrian Biofuels Institute)

extracted from the fleshy part of the fruit, which contains 45 to 55 percent oil, and it is used mainly in the manufacture of soaps, candles, margarine, and cooking oils. Palm kernel oil, as its name implies, comes from the kernel of the fruit, which contains about 50 percent oil. Nearly colorless, palm kernel oil is solid at room temperature and is used in making ice cream, mayonnaise, baked goods, and soaps and detergents. The pressed "cake" (pulp) remaining after the oil has been extracted is used as an animal feed. The African palm can be found along the coast of West Africa from Liberia to Angola and eastward to the Indian Ocean islands of Zanzibar and Madagascar. The African palm is also sometimes grown as an ornamental tree in other subtropical locations, such as Florida or Southern California.[3] Palm oil is the main biodiesel feedstock in Malaysia, but palm oil biodiesel has a high *cloud point* (the temperature at which the first wax crystals appear in biodiesel), making it less desirable for use in colder climates in its pure form. Despite its top ranking in terms of oil yield, palm oil at present represents only about 1 percent of total global biodiesel raw material sources.[4]

Coconut

As a source of oil, the coconut tree produces only around half as much oil as the oil palm, which is still an impressive 2,689 liters per hectare (287 gallons per acre) annually. Coconut is the third-most-produced oil in the world, after peanut and soybean. To make coconut oil, the meat of the coconut is peeled from the husk, dried, and pressed (or the meat can be shredded fresh and cold-pressed). The residual cake is used as an animal feed. Among the many products made from various parts of the coconut tree are twine or rope, mattress padding, mats, rugs, brushes, filler for plastics, charcoal filters, vinegar, soaps, lubricants, hydraulic fluid, paints, synthetic rubber, and margarine; it can also be used in making ice cream. The coconut palm may have originated in northwestern South America, but it is now found in all tropical regions of the world, especially along coastlines, and increasingly in plantations (which may have some negative environmental implications in certain locations).

Jatropha

Jatropha is a versatile bush or tree that has many uses. It has garnered increased interest recently because it adapts well to semi-arid, marginal locations and can be grown as a hedge for erosion control, property boundaries, and animal fencing. Jatropha is used to make lamp oil, soap, candles, poisons, and a wide range of folk remedies. Jatropha can produce about 1,590 liters of oil per hectare (202 gallons per acre) annually. Widely grown as a medicinal plant, jatropha establishes itself easily and is found in Brazil, Fiji, Honduras, India, Jamaica, Nicaragua, Panama, Puerto Rico, Mexico, El Salvador, and much of Africa. Jatropha is presently a very minor source of biodiesel, but it has a lot of potential.

Rapeseed/Canola

Sometimes cultivated in small quantities as a potherb, this yellow-flowering plant is more commonly grown in Europe as a forage feed for livestock and as a source of rapeseed oil (known in North America as canola oil). Rapeseed/canola produces about 1,190 liters per hectare (127 gallons per acre), giving it the highest yield of any conventional oilseed field crop. The residual seed is used as a high-protein animal feed. Rapeseed is cultivated in most European countries, Canada, and Russia, but it will grow in most temperate regions. Rapeseed, which represents 84 percent of global biodiesel raw material sources, is the principal feedstock for biodiesel produced in Europe, and most biodiesel research is based on rapeseed methyl ester or rapeseed ethyl ester.

Peanut

Peanuts produce an edible oil that can be used for cooking and deep-frying; in margarines, salad dressings, and shortenings for pastry and breads; and for the manufacture of pharmaceuticals, soaps, and lubricants. The seeds (peanuts) are also eaten raw, roasted and salted, chopped in confectioneries, or (especially in the United States) ground into peanut butter. In most other countries, peanuts are mainly processed for oil. The peanut produces about 1,059 liters per hectare (113 gallons per acre). Native to South America, peanuts are now widely cultivated in warm climates and sandy soils throughout the world. Although one of Rudolf Diesel's engines used peanut oil as fuel at the 1900 Paris Exposition, today it is a relatively minor source of feedstock for biodiesel.

Sunflower

The sunflower is cultivated mainly for its seeds, which yield the world's second-most-important type of edible oil, and it produces around 952 liters per hectare (102 gallons per acre). Sunflower oil is used for cooking, in margarine and salad dressings, and in lubricants, soaps, lamp oil, and a variety of paints and varnishes. Sunflower kernels are eaten raw or roasted and salted; they also can be made into flour. The cake (with seed hulls removed) is used as a high-protein animal feed. Native to western North America, the sunflower is the only major oil crop to have evolved in what is now the United States. Introduced early to Europe and Russia, sunflowers now grow in many countries, in both temperate and tropical regions. Sunflowers represent 13 percent of global biodiesel raw material sources.

Safflower

An annual thistlelike herb, safflower is grown mainly for its edible-oil-producing seeds, which yield about 779 liters per hectare (83 gallons per acre) of safflower oil. High in essential unsaturated fatty acids, safflower oil is light colored and easily clarified. It is used in salad and cooking oils, margarine, and candles and as a drying oil in paints, linoleum, and varnishes. Safflower is believed to have originated in southern Asia and was cultivated in China, India, Persia, and Egypt from prehistoric times. Later, safflower was grown in Europe and was subsequently introduced to Mexico, South America, and the United States. Safflower, which does well in the same areas that favor the growth of wheat and barley, can be planted, cultivated, and harvested with standard farm machinery used for small grains. Safflower oil is not a major biodiesel feedstock at the present time.

Mustard

Several varieties of mustard are grown as vegetable greens and as cover crops, but they are especially valued for their seeds, which can produce about 572 liters per hectare (61 gallons per acre) of mustard oil. The seeds can also be processed into a wide range of products, including various types of commercial mustards, lubricants, and hair oil. The seed residue is fed to animals, used in fertilizers, or used to make a valuable organic pesticide. Widely grown in many countries, mustard is considered a weed on cultivated lands in some locations. Most mustard crops can be planted, cultivated, and harvested using ordinary farm machinery. Although mustard seed oil is not widely used as a biodiesel feedstock, quite a lot of research is presently focused on its use, particularly in the United States.

Soybean

The soybean produces one of the world's most important sources of oil and protein. Soybeans yield around 446 liters per hectare (48 gallons per acre) of edible oil, which is used as a salad oil and for the manufacture of margarine and shortening. The oil is also used industrially in the manufacture of paints, linoleum, printing inks, soap, insecticides, disinfectants, and a wide range of other products. The most commonly raised crop in the United States (around 2.9 billion bushels were produced in 2004), the soybean has been grown in East Asia for thousands of years and is widely cultivated in subtropical and tropical regions. Although most biodiesel in the United States is made from soybean oil (virgin or used), soybeans are not the best crop for oil production. This is reflected worldwide, where soybeans represent only 1 percent of total global biodiesel raw material sources.

Corn (Maize)

Corn (also known as maize) is believed to have originated in Mexico in prehistoric times. Few plants are grown as extensively or put to more uses. Corn is a staple cereal food in Central and South America and throughout much of Africa. The oil produced, which amounts to only about 172 liters per hectare (18 gallons per acre), represents about 7 to 8 percent of the grain, placing corn at the bottom of the oil-production spectrum. Nevertheless, processed corn oil is used extensively as a frying oil in the United States. Ethanol, a renewable alcohol fuel, is made from cornstarch in the United States.

Used Cooking Oil

The United States is almost as addicted to fast food as it is to petroleum. Every year, over 3 billion gallons (11.3 billion liters) of used cooking oil is drained from deep-fat fryers in this country. Most restaurants and other frying-oil users have to pay a rendering or waste company to haul the greasy stuff away. Some used cooking oils are processed into animal feed, makeup, or fertilizer, but much of it has been dumped in landfills or sewers in some locations, causing a lot of headaches for sewage treatment plants. But used vegetable oil is a great low-cost resource for making biodiesel, and it is increasingly used for that purpose in the United States, and to a lesser extent in Europe and elsewhere. However, because it is of lower quality due to repeated heating at high temperatures and possible contamination with animal fats and food particles, used cooking oil requires additional treatment, both before and after the transesterification process.

This is an important issue if the biodiesel made from it is to meet stringent government standards for use as a vehicle fuel. Despite this fact, turning what would otherwise be a waste product into usable fuel makes a lot of sense from a variety of environmental perspectives. And in locations where large quantities of used cooking oil are available, such as in major cities, there is a lot of potential for making substantial amounts of biodiesel. For example, according to the *New York Times*, it would take just one-fifth of the waste cooking oil produced in New York City to run its entire public transit bus system.

Animal Fat

Animal fat, a by-product of the animal rendering process, is the least expensive feedstock currently available for biodiesel production. Beef tallow, poultry fat, fish oil, and other types of animal fats will all work. In recent years, the United States has generated around 11 billion pounds of animal fats, according to U.S. Department of Agriculture averages, which theoretically is enough to produce roughly 1.5 billion gallons of biodiesel. Substantial quantities of animal fats are available in many other countries for potential biodiesel production as well. Admittedly, not all of this feedstock is available for biodiesel production, but it does present an opportunity to lower input costs substantially.

However, biodiesel made from animal fats tends to have poor cold-weather properties compared with biodiesel made from virgin oils such as soybean or rapeseed. Nevertheless, in warmer climates around the world, there is considerable potential for using animal fats as a feedstock. What's more, when biodiesel is made from animal fats, it generally has lower tailpipe nitrous oxide (NOx) emissions than other biodiesels. The growing global concern about how to dispose of cattle infected with bovine spongiform encephalopathy (BSE, or mad cow disease) or otherwise suspect cattle has provided another opportunity for the biodiesel industry, especially in the European Union, where feeding ground-up animal parts to other animals has been banned since January 2001. As a result, a market for the production of biodiesel from rendered animal fats has begun to develop in a number of European nations (see chapter 6).

Algae

There is one additional biodiesel feedstock that potentially dwarfs all the others in terms of oil production: algae. While pond algae may seem like a bizarre source of diesel fuel, it's not as far-fetched as it may sound. That's because much of the original organic matter that formed the basis of the world's present petroleum resources was algae—vast amounts of algae—in shallow, prehistoric bodies of water. These deposits were subsequently buried by sediments and later transformed by pressure and heat over millions of years into petroleum. Today the process can be shortened to just a few days simply by growing the algae in ponds of relatively saline (salty) water and extracting the oil directly from the harvested algae.

From 1976 to 1998, the National Renewable Energy Laboratory (NREL) conducted a $25 million program funded by the U.S. Department of Energy called the Aquatic Species Program: Biodiesel from Algae. The main focus of the program was the production of biodiesel from algae grown in ponds utilizing waste CO_2 from coal-fired power plants. While the program's use of one of the most environmentally unfriendly fossil fuels—coal—in its experiments seems incongruous, it was at least utilizing the carbon dioxide emissions in a productive way (this is very much in line with the current idea of "waste exchange," where the waste from one industry is used as a feedstock for another).

One of the main accomplishments of the program was to locate, catalog, and study over three thousand strains of algae from all across the continental United States and Hawaii. After preliminary testing in California and Hawaii, a large-scale pond facility was built in the desert near Roswell, New Mexico. The quantity of oil that the algae in Roswell could produce was remarkable. A series of 1,000-square-meter ponds were filled with floating algae. Up to 50 grams of algae were produced per square meter daily.[5] Thus, each pond had a potential daily yield of 50,000 grams (50 kilograms, or 110 pounds). Oil extraction rates between 30 and 40 percent are not unusual for some types of algae, so using a conservative 30 percent as a multiplier yields approximately 15 kilograms (33 pounds) of oil per pond per day. This means that one pond could theoretically produce 5,475 kilograms (12,045 pounds) of oil per year. Assuming that 7.5 pounds (roughly) equals a gallon, one algae pond could possibly yield 3,418 liters (roughly 903 gallons) of oil per year (or about 3,654 gallons per acre per year). However, in actual practice, the Roswell researchers were unable to maintain consistent production levels due to cold overnight desert temperatures. Other researchers have estimated that one acre of algae could produce anywhere from 500 to 20,000 gallons of algal oil every year, so there is clearly a very wide range of opinion on this issue.

The NREL researchers concluded that "algal biodiesel could easily supply several 'quads' of biodiesel—substantially more than existing oilseed crops could provide." (A quad represents one quadrillion Btu of energy.) Two hundred thousand hectares (494,000 acres), representing less than 0.1 percent of suitable land area in the United States, could produce one quad of fuel.[6] Two quads would represent an area the size of the state of Rhode Island, the smallest state in the union. Regardless of whose production statistics you use, figures like this clearly demonstrate the potential for huge quantities of oil production from algae. Although the NREL tests were reluctantly terminated in 1998 due to budget cuts, the researchers hoped that their work would be used as a foundation for the future production of biodiesel from algae.

Production Potential

Total global output of biodiesel is approaching 2 million metric tons (570 million gallons) annually. About two-thirds of that capacity is generated by western Europe, with eastern Europe, the United States, and the rest of the world making up the balance. By comparison, the United States alone consumes approximately 58 *billion* gallons of middle-distillate fuels annually, according to the U.S. Energy Information Agency (middle-distillate fuels include diesel fuel, heating oil, kerosene, jet fuels, and gas turbine engine fuels). Of that total, about 35 billion gallons (60 percent) is for on-highway vehicle use, while the rest is divided among a wide range of residential, commercial, and industrial purposes. How much of this consumption can be met by biodiesel? This is a question that has sparked quite a lot of debate. As noted earlier, the United States generates about 3 billion gallons (11.3 billion liters) of used frying oil every year. Assuming that half of this oil could be converted to biodiesel, that would represent perhaps 2.5 percent of the current petrodiesel market. If half of all 11 billion pounds of animal fats produced in the United States could be processed into biodiesel, that might yield roughly 750 million gallons of biodiesel, or 1.25 percent of the petrodiesel market. If all the fallow cropland in the United States (about 60 million acres) were planted to rapeseed (admittedly a large assumption) and yielded 100 gallons of oil per acre (not such a large assumption), that could produce roughly another 6 billion gallons, or about 10 percent of the current petrodiesel market. The total estimated production from all of these sources comes to 8.25 billion gallons, or about 14 percent of the U.S. petrodiesel market. All of these sources could be used without displacing any current food crop production, an important consideration. This projection admittedly does not take into account possible biodiesel production from algae, which, at the present time is not occurring on a large scale and would require massive investment in production facilities. Optimistically, several quads of production could be added for algae, which would boost total U.S. biodiesel production by about 20 billion gallons to roughly 28 billion, approximately half the present petrodiesel market. Although production capacities in other countries vary quite a bit, in Europe it is estimated that biodiesel has the potential to replace somewhere between 10 and 15 percent of the current petrodiesel market.[7]

How do these admittedly sketchy (and optimistic) estimates compare with "official" projections for the United States? In terms of production facilities, the National Biodiesel Board of Jefferson City, Missouri, says that two separate figures need to be considered when looking at biodiesel production capacity: dedicated capacity and available capacity. *Dedicated capacity* refers to facilities that are specifically designed to make biodiesel and can produce between 60 and 80 million gallons per year (the actual production for 2003 was only about 25 million gallons). Much of this production capacity is modular, which means that it could easily be doubled or tripled in less than a year. *Available capacity* refers to facilities in the oleochemical industry that make fuel-grade methyl esters as a normal part of their chemical businesses. The NBB estimates that there currently is about a 200-million-gallon excess capacity in that industry that could be used to produce biodiesel.

Looking beyond production facilities, the National Renewable Energy Laboratory estimates that around 1 billion gallons of biodiesel could be produced in a year or so in the United States, and that production could be boosted to 2 billion gallons by 2010. Another 4 to 10 billion gallons might be produced using mustard seed, according to the Department of Energy Biofuels Program (now the Biomass Program). These official estimates come to a total of 12 billion gallons, or about 20 percent of the present U.S. petrodiesel market. This is a very optimistic projection. Whether the correct figure is 14 percent or an even more optimistic 20 percent, it's clear that future U.S. biodiesel production capacity (in terms of both facilities and feedstocks) falls far short of the current petrodiesel market, unless algae were used as a biodiesel feedstock in a massive, industrial-scale initiative that would take many years and a lot of money to develop. It would be difficult, but not impossible. Especially if a *real* national energy plan (rather than the usual handouts to the fossil fuel and nuclear industries), one that strongly emphasized renewables, was finally implemented in the United States. Also difficult, but, again, not impossible.

How Does Biodiesel Compare?

By now it should be obvious that biodiesel has a lot going for it. It can be made almost anywhere by almost anybody from a wide range of ingredients. But how does it really stack up when compared to petrodiesel? In most respects biodiesel is better, while in others it's not quite as good. First, biodiesel can be used in any modern diesel engine without any modifications to the engine (this is not the case with some other fuels, such as compressed natural gas, which require major capital expenditures for engine retrofits or new engines as well as new infrastructure). Biodiesel has excellent lubricating properties and will lubricate many moving parts in the engine, increasing engine life. It also has a higher *cetane number* than petrodiesel (cetane is a measure of the ignition quality of diesel fuel), indicating better ignition properties. On the downside, the energy content of biodiesel is about 10 to 12 percent lower than that of petrodiesel (about 121,000 Btu compared to 135,000 Btu for Number 2 diesel fuel). Biodiesel has an oxygen content of around 10 percent, more than that of petrodiesel fuel. The oxygen results in more favorable emission levels but also results in reduced energy content. Although this causes a slight reduction in engine performance, the loss is partly offset by a 7 percent average increase in combustion efficiency. Generally speaking, the use of biodiesel results in about a 5 percent decrease in torque, power, and fuel efficiency in diesel engines.[8] However, most people can't detect any noticeable difference in the performance of their vehicles.

Biodiesel is free of lead, contains virtually no sulfur or aromatics (toxic compounds such as benzene, toluene, and xylene), and results in substantial reductions in the release of unburned hydrocarbons, carbon monoxide, and particulate matter (soot), which has been linked to respiratory disease, cancer, and other adverse health effects. The production and use of biodiesel results in a 78 percent reduction in carbon dioxide emissions, according to a joint U.S. Department of Energy (DOE) and U.S. Department of Agriculture (USDA) study published in 1998.[9] This is mainly because the plants used to grow biodiesel feedstock absorb most of the CO_2 emissions from biodiesel combustion. On the downside, emissions of nitrogen oxides (NOx, a contributing factor in the formation of smog and ozone) are usually slightly greater with biodiesel. This can be partly offset by tuning the engine specifically for biodiesel. Recent additive tests show promising results for reducing nitrogen oxide emissions as well. Last but not least, biodiesel replaces the typically noxious exhaust smell of petrodiesel with an odor faintly like that of french fries or doughnuts, especially if it has been made from recycled cooking oil.

Life-Cycle Studies

Biodiesel has other positive features as well. The overall life-cycle production of wastewater from the production of biodiesel is 79 percent lower than the overall production of wastewater from petrodiesel. What's more, the overall life-cycle production of hazardous solid wastes from biodiesel is 96 percent lower than overall production of hazardous solid wastes from diesel, according to the DOE/USDA study.

Another important factor when comparing biodiesel with other fuels is what is known as the *energy efficiency ratio*. This is a numerical figure that represents the energy stored in the fuel compared to the total energy required to produce, manufacture, transport, and distribute the fuel. The total energy efficiency ratios for biodiesel and petrodiesel are very similar. However, the total *fossil* energy efficiency ratio for biodiesel shows that biodiesel is about four times as efficient as petrodiesel in utilizing fossil energy, according to the DOE/USDA study cited above. Biodiesel has a positive fossil energy efficiency ratio of 3.2 to 1. Petrodiesel, on the other hand, has a negative fossil energy efficiency ratio of 0.83 to 1, according to the study.[10] Just for the sake of comparison, ethanol has an energy efficiency ratio of 1.34 to 1, according to a 2001 USDA study.[11] Other studies say that biodiesel has an energy efficiency ratio of around 2.5 to 1, while still others say it can be as high as 7 to 1. Energy efficiency ratios and similar statistics are tricky and can vary widely depending on what factors are included in the studies, so it's not unusual to see a wide range of figures. In any case, biodiesel generally provides more energy with far less negative environmental impacts on a life-cycle basis than petrodiesel.

Biodegradability and Toxicity

When it comes to biodegradability and toxicity, biodiesel wins the contest over petrodiesel hands down. Biodiesel fuel is not harmful to humans or the environment. One hundred percent biodiesel (also called B100 or "neat" biodiesel) is as biodegradable as sugar and up to ten times less toxic than table salt. One study showed that biodiesel rubbed on the skin was less irritating than a 4 percent water and soap solution. While biodiesel smells somewhat, well, greasy, otherwise it's relatively benign. On the other hand, just the fumes from petrodiesel are toxic and dangerous.

While ingesting carefully measured amounts of biodiesel (not recommended) caused no major problems for a group of laboratory rats in a 1996 Ohio laboratory study, drinking even a small amount of petrodiesel could prove extremely dangerous, even fatal for humans (or rats). Studies have also shown that if spilled accidentally, biodiesel will biodegrade up to four times faster than petrodiesel fuel. In about three to four weeks' time, biodiesel will be almost entirely biodegraded.[12] This is an especially important advantage in environmentally sensitive areas such as national parks and forests or waterways. While accidental spills of biodiesel will still cause some environmental problems, they are minimal when compared to an equivalent spill of petroleum diesel or crude oil.

Transportation and Storage

Biodiesel is much safer to transport and store than petrodiesel. That's because, aside from its low toxicity, biodiesel also has a high *flash point*, or ignition temperature. This means that biodiesel needs to be above 260°F (126°C) before it will ignite. The flash point of petrodiesel, by comparison, is around 125°F (52°C), which is one reason why it is considered a hazardous material. Biodiesel, on the other hand, is so safe that it can be shipped by common carrier, or even UPS or FedEx. Tests have also shown that the flash point of biodiesel blends increases as the percentage of biodiesel increases. Consequently, biodiesel and blends of biodiesel with petrodiesel are safer to ship and store than straight petrodiesel.

In general, the same storage and handling procedures used for petrodiesel can be used for biodiesel, meaning that no major changes to infrastructure are needed. Biodiesel should be stored in a clean, dry, dark environment in tanks made from steel, aluminum, fluorinated polyethylene, fluorinated polypropylene, or Teflon. Copper, brass, tin, lead, and zinc should be avoided, according to the National Biodiesel Board.

Due to its organic nature, biodiesel is somewhat more susceptible to the growth of bacteria and mold than petrodiesel. Consequently, it is sometimes necessary to add small quantities of biocides to stored biodiesel, especially in warmer climates, to reduce the growth of these contaminating organisms. Ideally, the storage time for biodiesel (as with petrodiesel) should be limited to six months.

Biodiesel Blends

In addition to its many environmental advantages, another extremely useful characteristic of biodiesel is that it can be blended with petrodiesel in any percentage. A B20 blend (20 percent biodiesel and 80 percent petrodiesel) has demonstrated significant environmental benefits with a minimum increase in cost to consumers. In recent years, B20 has become a popular fuel with many fleets in the United States, encouraged in part by the 1998 Congressional approval of B20 as a compliance strategy for fleets under the Energy Policy Act of 1992 (for more on the EPAct, see chapter 10). It is biodiesel's ability to be blended with petrodiesel at any percentage from B1 to B99 that makes it such a flexible fuel that can meet a wide variety of different needs, as we will see in the next chapter.

4

Biodiesel's Many Uses

Although Rudolf Diesel's early designs were for large, industrial-scale engines in stationary settings, he believed that his invention would eventually be used to power a wide range of vehicles of virtually every description. His vision eventually came to pass (after his death), but by then his engine design had been modified to burn petroleum-based diesel fuel. There were, however, a few exceptions. In the 1950s, Germany's MAN AG developed its "salad oil engine" design (named for its ability to run on a wide variety of different fuels), but the design was mainly confined to limited numbers of larger stationary installations.[1] The vast majority of diesel engines in use around the world, however, burned (and still burn) petrodiesel fuel, which is similar to kerosene, jet fuel, and home heating oil. Consequently, early biodiesel researchers were focused on creating a renewable fuel for diesel engines that were optimized to run on petrodiesel. As we have seen, there are many feedstocks that can be used to make biodiesel. But the different ways of producing biodiesel are exceeded by the many ways of using it. In this chapter we'll explore those applications. First, though, we'll take a quick look at the modern diesel engine and a few issues related to using biodiesel as a fuel.

How the Diesel Engine Works

Although there are many variations in size and usage, most diesel engines share the same basic features they inherited from Rudolf Diesel's original designs. Like their gasoline-powered relatives, diesels have a number of cylinders (usually four, six, or eight) containing pistons that are connected to a crankshaft. When the pistons move up and down in the cylinders, they cause the crankshaft to turn. The crankshaft's rotational force passes through a transmission to a drive shaft that ultimately turns the wheels of the vehicle (or, in the case of ships, a propeller). One of the main differences between a spark-ignition (gasoline) engine and a compression-ignition (diesel) engine is in the fuel system and how the fuel is combusted. In a gasoline-powered engine the spark plug causes the fuel to ignite. In a diesel engine the extreme high pressure and temperature (around 1,000°F/540°C) of the air in the top of the cylinder causes the fuel to ignite spontaneously, without the use of an electric spark. One key advantage of diesel engines, therefore, is that they don't need spark plugs, an ignition coil, a distributor, or a carburetor. Most diesel engines do, however, need what is called a "glow plug," which preheats the combustion chamber and helps the fuel ignite when the engine is being started. Because a diesel engine compresses air at a much greater force (with a compression ratio of as much as 24 to 1), its engine block and many other internal components are stronger than in a gasoline-powered engine. This results in a heavier, more expensive, but longer-lasting engine. It is not unusual for diesel vehicles to run for 200,000 to 300,000 miles or more before needing an engine overhaul.

Although there have been many advances in diesel engine design, some of the most significant progress over the years has been related to improvements in fuel-injection technology. Germany's Robert Bosch is generally credited with developing the world's first commercial diesel-injection pump. Prototypes were tested as early as 1923, and the pump was mass-produced beginning in 1927.[2] The pump was the key development that made it possible to produce smaller, lighter diesel engines for use in trucks and automobiles. The fuel system on a diesel engine consists of a fuel tank, an injector pump, and the injectors, which, as their name implies, inject fuel into the engine under high pressure. Older diesel engines use what is called indirect injection, in which the fuel is injected into a prechamber, where it is partly combusted before it enters the cylinder. Most newer diesels use what is called direct injection, or turbo direct injection (TDI), in which the fuel is injected directly into the cylinder. The older engines are tough and extremely reliable but noisier, and they tend to emit more soot (partially burned fuel). The newer, direct-injection engines are quieter and cleaner and have better acceleration. Modern diesel engines come in a wide range of sizes and designs for many different uses, but their rugged dependability makes them the motive power of choice for heavy commercial and industrial uses and other applications.

Cold-Weather Issues

Some diesel engines are notoriously hard to start in cold weather. The reason for this is because petrodiesel begins to "cloud" at about 20°F (-7°C), the temperature at which paraffin wax crystals begin to form. Cloudy diesel fuel can clog fuel filters and keep the engine from starting or cause it to stall. When temperatures fall further, diesel fuel reaches its *pour point* (the temperature below which it will not pour). At this point diesel fuel generally stops flowing through fuel lines and diesel engines stop running. Normally the cloud point and the pour point are about 15 to 20 degrees Fahrenheit apart. When the temperature drops even further, to about 15°F (-9.5°C), diesel fuel reaches its *gel point*, when it becomes the consistency of petroleum jelly. Biodiesel, unfortunately, suffers from all of these problems, but at higher temperatures, making the situation even worse than it is with petrodiesel. The actual cloud point for biodiesel varies depending on what kind of material was used for the feedstock. Biodiesel made from used cooking oil or animal fats will cloud at higher temperatures than biodiesel made from virgin rapeseed/canola oil. But there are even different cloud points among various types of virgin oils. As noted in chapter 3, palm oil biodiesel has a high cloud point and generally is not suitable for use in extremely cold climates.

During the winter months in cold regions, petrodiesel fuel is normally altered with special winter formulations to help it perform better. There are also winterizing agents, antigel formulas, and other additives that can lower the cloud point of petrodiesel. Fortunately, these same agents and formulas can be added to biodiesel blends to improve their winter performance as well. Another widely used strategy is to use a lower biodiesel blend, such as B20, during the winter and a higher biodiesel concentration, such as B50 or B100, during the warmer months. For hard-core biodiesel purists who live in cold climates, special cold-weather heating kits are available for most diesel cars and trucks.

Sludge and Slime

One characteristic of biodiesel that can cause some problems in older diesel engines is its high solvent potential. On engines manufactured before 1994, the rubber seals, hoses, and gaskets will be degraded with the use of high concentrations of biodiesel, especially B100. If these rubber parts are replaced with biodiesel-resistant materials such as Viton, B100 can be used without any problems. But biodiesel's solvent properties can also cause problems with old fuel tanks and fuel lines, which are typically coated with sludge. The biodiesel dissolves the sludge, which then can end up in the fuel filter, causing the engine to malfunction. However, once the sludge has been cleaned out of the fuel system, the engine should run without further trouble on biodiesel.

In warm climates, diesel engines—and especially their fuel tanks—are susceptible to bacteria growth, which can clog the fuel system and cause engine failure. This problem can occur with either petrodiesel or biodiesel fuels. The typical greenish to black bacterial slime grows in the absence of light but in the presence of moisture in the fuel tank as it feeds on the hydrocarbons in the fuel. The bacteria can be eliminated with the use of biocides, which are widely available at automotive parts stores, fuel dealers, and other retail outlets. Keeping the fuel tank as full as possible will minimize the amount of condensation, oxygen, and bacteriological activity.

Engine Warranties

Although biodiesel will run in virtually any diesel engine, not all engine manufacturers will honor their engine warranty if biodiesel has been used as a fuel. Today this is less of a problem than it used to be, because more experience in the use of biodiesel has been gained. In the past, the main problem with using biodiesel was its solvent properties, which softened rubber engine parts. Most diesel engines manufactured after 1994 use components that are biodiesel-resistant. In the United States, diesel engines are manufactured for petroleum diesel fuels that meet the requirements set by the American Society for Testing and Materials (ASTM). For petrodiesel fuel, the ASTM standard is ASTM D 975. In 2001 ASTM approved a new standard for biodiesel, D 6751, covering pure biodiesel (B100) for blending with petrodiesel up to 20 percent by volume (B20). Many major U.S. diesel engine manufacturers now say that the use of blends up to B20 that meet the ASTM standard will not void their warranties. This also includes blends below B20. For blends above B20, the individual engine manufacturer should be contacted.

Diesel engine manufacturers guarantee their engines against defects in "materials and workmanship." They do not warrant the fuels. Consequently, if there is a fuel-related problem with an engine (whether it is petrodiesel or biodiesel), the fuel manufacturer is responsible. This is why it is important to purchase biodiesel fuel from a commercial supplier who will certify that its product meets the ASTM standard. The National Biodiesel Board has formed the National Biodiesel Accreditation Commission (NBAC) to support biodiesel fuel quality standards in the United States. The NBAC issues a "Certified Biodiesel Marketer" seal of approval to biodiesel marketers that have met its standards. People who make their own biodiesel have only themselves to blame if they damage their engines with off-spec fuel. This seems to be a risk most of them are willing to take, but it is something that large fleet operators simply cannot afford to do.

In Europe, biodiesel standards have been set by individual countries for many years. Austria's first standard was issued in 1991 and was the basis for numerous engine warranties. A German standard, DIN 51606, was formalized in 1997 and has been widely used as a guide for other European national standards. After many years of effort, a new European-wide biodiesel standard, DIN EN 14214, was published in October 2003 that essentially supersedes previous European national standards as of March 2004. Many European engine manufacturers' warranties cover the use of biodiesel.

Where the Diesel Engine Is Used

Now it's time to take a look at where diesel engines are used today. The transportation sector is responsible for more than 70 percent of the petroleum consumed in the United States and one-third of U.S. carbon dioxide emissions. The same general statistics apply to Europe as well. Over-the-road vehicles account for the vast majority of diesel fuel use (34 billion gallons or 59 percent of middle-distillate fuels) in the United States. When most Americans think of transportation they generally think first of automobiles. But the transport sector extends well beyond cars to include trucks, buses, trains, boats, and planes. And the vast majority of these other types of transport are powered by diesel engines or, in the case of commercial aircraft, on diesel-like aviation fuels. It's no exaggeration to say that the vast majority of the world's heavy transport sector is diesel powered. This offers obvious potential for a wide range of uses for biodiesel. Here is a brief overview of those applications.

Automobiles

Diesel-powered cars were first introduced to the European market in 1936, and their use in most European countries has grown steadily since. Diesel cars became popular as alternative-fueled vehicles in the United States during the oil crises of the 1970s. The main reasons for this were that diesel vehicles got better mileage and diesel fuel was relatively inexpensive in those days (the fact that diesel was still a petroleum-based fuel seems to have escaped the attention of a lot of people initially). But most American drivers were not impressed by the slower acceleration and noisier operation of diesel automobiles, some of which were poorly designed. By the mid-1980s, demand for diesel cars fell off, and production declined as well. With the return of cheap gasoline after the oil crisis ended, U.S. drivers went back to driving their gas-guzzlers. Meanwhile, much of the rest of the world continued to drive diesel-powered cars and light trucks, due mainly to higher fuel prices. In the late 1990s, diesel cars and pickup trucks returned to the U.S. market with new TDI engines. Volkswagen and Mercedes-Benz both offered diesel-powered cars, while Dodge and Ford offered diesel-powered pickups. Despite the reintroduction of diesels to the United States, they represent only about 1 percent of total pleasure cars driven in this country, while in Europe they represent about 36 percent. Diesel cars in Europe are increasingly popular and are expected to account for about 50 percent of the market by 2006. Using biodiesel as a fuel for automobiles in countries that have larger percentages of diesel-powered cars obviously has more potential for reducing petrodiesel fuel consumption in the automotive sector than in the United States.

Fleets

In the United States, biodiesel is being used in over five hundred fleets across the country, and that number is growing rapidly. All four branches of the U.S. military, the Postal Service, and dozens of school districts and municipal fleets have been jumping on the biodiesel bandwagon in recent years. The Postal Service alone used half a million gallons of biodiesel fuel in 2002. In California, the city of Berkeley runs its trucks on B100. The city of Boulder, Colorado, has been testing biodiesel in its tractors, dump trucks, and fire trucks. In 2000 Lambert International Airport in Saint Louis switched to a B20 blend in three hundred of its vehicles, including deicer tankers, snowplows, high-speed runway brooms, lawn equipment, passenger shuttles, and aircraft rescue fire trucks. "The reliability is great," says Frank Williams, fleet maintenance foreman. "We've had sustained 25-below wind chill factor for multiple days and never had a problem with the B20. Most of the vehicle operators didn't even know we had switched to biodiesel."[3] Unlike compressed natural gas, biodiesel does not require expensive engine retrofits, making the switch to biodiesel attractive to large fleet owners. And since many municipalities across the United States have been suffering from severe budget limitations in recent years, they can keep their older vehicles running longer while lowering emissions at the same time by using biodiesel.

Mass Transit

Some of the most highly visible users of diesel engines are mass transit fleets. Diesel engines power nearly 90 percent of the more than seventy-six thousand active transit buses in the United States, according to the American Public Transportation Association. Since most large urban bus systems typically have hundreds or even thousands of buses on the road, they can be significant contributors to urban air pollution. An increasing number of transit fleets are reporting positive experiences with biodiesel. Among the many city bus fleets in the United States using biodiesel are those in Cedar Rapids, Iowa; Bloomington, Indiana; Saint Louis, Missouri; Oklahoma City, Oklahoma; Olympia and Seattle, Washington; Raleigh, North Carolina; and Springfield, Illinois.

One of the most successful biodiesel initiatives in a city transit fleet has taken place in Graz, Austria. Beginning in 1994, the Grazer Verkehrsbetriebe (GVB) began field tests with two of its city buses running on biodiesel made from recycled frying oil. After many years of continued positive test results, the GVB plans to convert its entire fleet to biodiesel by 2005.[4] Many other bus fleets around the world are experimenting with (or regularly using) biodiesel. In Italy, public transit buses in Florence, Gorgonzola, Padua, and Perugia now run on biodiesel. French buses in Paris, Bordeaux, Dijon, Dunkirk, Grenoble, and Strasbourg are using a biodiesel blend. Biodiesel systems in Canada include those of Brampton, Ontario, and Saskatoon, Saskatchewan. The use of biodiesel in urban transit systems around the world is growing exponentially.

School buses are one of the largest mass transit programs in the United States. Approximately 460,000 school buses transport more than 24 million children to and from schools and school-related activities every school day.[5] The vast majority of these buses are powered by diesel engines. The use of biodiesel in school bus fleets is growing not only due to environmental awareness but also because of health concerns about students. Several thousand school buses are currently operating successfully on biodiesel blends in the United States, including those in school districts in Olympia, Illinois; Clark County, Nevada; Denver, Colorado; and Medford, New Jersey. "It's been absolutely fantastic," said Joe Biluck Jr., director of operations and technology for the Medford district. "We've had no down time as a result of this fuel. We've never had a fuel system gel up on us and we've run down to temperatures of 11 degrees below zero and haven't experienced any problems."[6] Biodiesel-powered school buses can be found throughout many parts of Europe as well.

Trucks and Heavy Equipment

Commercial trucks and heavy industrial equipment are almost entirely powered by diesel engines. The use of biodiesel in heavy construction and mining equipment has been growing slowly in the United States in recent years. Rockland Materials, a Phoenix, Arizona, concrete company, switched to biodiesel in 2001. The company has a fleet of about one hundred diesel ready-mix trucks, dump trucks, and semitrailers. Carmeuse Lime Mines, which operates two of the nation's largest lime mines in Maysville and Butler, Kentucky, runs more than 150 pieces of underground equipment on biodiesel and is now the largest single user of biodiesel in the state. In 2002 Alcoa Davenport Works in Davenport, Iowa, switched all of its diesel-powered equipment to a B20 blend. This aluminum manufacturing plant used approximately 250,000 gallons of biodiesel in 2003, roughly the equivalent of all the soybeans from a 100-acre farm, according to the National Biodiesel Board. However, except for a number of tests conducted by various research organizations, the use of biodiesel by the commercial, over-the-road trucking industry has not made much progress in the United States due to the highly competitive nature of the industry and the higher costs of biodiesel. If the cost of biodiesel could be lowered to make it competitive with that of petrodiesel, the trucking industry would almost certainly use more biodiesel because of its many other benefits, especially reduced maintenance costs and increased engine life. In Europe and elsewhere around the world, many trucking companies and heavy equipment operators are using biodiesel.

Farm Equipment

Most of the early biodiesel experiments were conducted on diesel farm tractor engines. The main idea was to guarantee the ability to grow food in the face of a future oil crisis. That rationale is just as valid today. But diesel engines power much more than tractors on most large modern farms. Biodiesel can be used as a fuel or fuel additive in virtually any diesel-powered farm equipment, such as trucks, harvesters, balers, irrigation pumps, and other machinery. This provides farmers with the opportunity to help create demand for biodiesel by using it themselves in their day-to-day operations while they are growing feedstock to make more biodiesel. Biodiesel blends offer improved lubricating properties over straight petrodiesel, as well as lower maintenance costs, less downtime, and increased equipment life.

One of the principal feedstocks for biodiesel in the United States is soybeans, a major crop produced by almost four hundred thousand farmers in twenty-nine states. A U.S. Department of Agriculture study completed in 2001 found that an average annual increase of the equivalent of 200 million gallons of soy-based biodiesel demand would boost total crop cash receipts by $5.2 billion cumulatively by 2010.[7] The farming sector itself in the United States consumes about 3.2 billion gallons of diesel fuel every year, representing about 5 percent of the middle-distillate fuel market, according to the Energy Information Administration. Biodiesel use in farming equipment around the world is growing steadily.

Boats

Rudolf Diesel's engine was adapted for marine use as early as 1903. Since then, diesel engines have spread to virtually every corner of the world's marine environments. Unfortunately, diesel engines can cause considerable environmental damage, especially in the case of a petrodiesel fuel spill. But it is precisely this environmental fragility that makes marine use of biodiesel so attractive. Tests have concluded that biodiesel is not harmful to fish, and that when spilled in water, biodiesel will be 95 percent degraded after twenty-eight days as compared with only 40 percent for petrodiesel in the same time period.[8] Diesel engines are used in a wide range of marine applications, including merchant ships, cruise ships, ferryboats, and powerboats, and even in sailboats as auxiliary engines. Electrical generators, bilge pumps, and other onboard equipment can also be diesel powered. One of the reasons petrodiesel fuel is popular in boats is due to its low risk of spontaneous combustion when compared with gasoline. Biodiesel, with its higher flash point, is even safer than petrodiesel.

Vessel operators report a noticeable improvement in the odor of engine exhaust when biodiesel is used instead of petrodiesel, making it less objectionable for engine crew and passengers alike. Anyone who has sat on deck downwind of the exhaust stack on a diesel-powered cruise ship knows that this is a less-than-relaxing experience. The marine industry presently accounts for about 10 percent of the petrodiesel consumed in the United States, so the potential for the increased use of biodiesel, especially in sensitive or protected waterway areas, is fairly substantial. Currently in the United States much of the emphasis on biodiesel is focused on recreational boats, which consume about 95 million gallons of diesel fuel every year, according to the National Biodiesel Board. However, the Washington State Ferries, one of the busiest and largest ferry systems in the United States, has been testing B5 and B10 biodiesel. Marine charter boat operators and other maritime businesses around the world are beginning to use biodiesel as well.

Trains

The testing of biodiesel as a fuel for diesel locomotives has been conducted on a limited basis in various countries. In December 2001 the Tri-County Commuter Rail Authority in southern Florida began running one of its locomotives on B100. The locomotive operated for three months in regular passenger service without any problems, and Tri-County continues to consider the future use of biodiesel. India Railways conducted its first trial run for B5 in a locomotive on the Delhi-Amritsar Shatabdi Express, one of its high-speed passenger trains, in December 2002. The railway plans to use locally produced biodiesel on more of its trains in the future. In December 2003 the transportation company América Latina Logística (ALL), with 15,000 kilometers of railroad in Argentina and southern Brazil, decided to replace about a quarter of the petrodiesel fuel it consumes with B20. Preliminary tests were reportedly conducted on two trains in early 2004, and there are plans to expand biodiesel use to the entire system.[9] There is a good deal of potential for additional expansion of biodiesel in the railroad industry worldwide, especially in congested urban areas where air quality is a particular concern.

Electrical Generators

Most larger electrical generators, whether for primary or standby generation, are powered by diesel engines. Even individual homeowners can use smaller diesel-powered generators in conjunction with off-grid energy systems or as emergency backup. Isolated communities in wilderness areas or on islands frequently rely on diesel generators as their sole source of electricity. Importing petrodiesel fuel for generator use in isolated locations can be very expensive (and potentially hazardous for sensitive environments). Large institutions such as hospitals, universities, military installations, and some businesses also use diesel-powered generators for emergency backup when grid power fails. All of these installations offer potential for expanded biodiesel use, especially in cities where backup generators can be significant sources of air pollution if they burn petrodiesel fuel.

Aircraft

In the United States, a number of studies have been conducted on the possible use of biodiesel as an aviation fuel for both military and civilian use. A 1995 study at Purdue University funded by the Indiana Soybean Growers Association found that biodiesel blended with jet fuel showed potential for use in aircraft with jet turbine engines, and that further testing was warranted.[10] The Renewable Aviation Fuel Development Center at Baylor University has also initiated research, development, and testing programs of blends of biodiesel and Jet A fuel for turbine engine aircraft. More research is needed, but results so far are encouraging.

Biodiesel as Lubricant and Solvent

Potential markets for biodiesel extend beyond the transportation and electrical-generation sectors. Even when used in low concentrations such as B2 or B5, biodiesel can offer a significant (up to 65 percent) lubricity advantage in any diesel engine. Biodiesel can even be used straight as a machinery lubricant. It is also possible to use biodiesel instead of kerosene in some camping lanterns and stoves. Biodiesel's solvent properties may be used to clean dirty or greasy engine or other machine parts; left in a bucket of B100, dirty parts are usually clean by the next morning.

But biodiesel can be used to clean more than dirty machinery. A series of laboratory experiments were conducted at the School of Ocean Science at the University of Wales to test the potential of biodiesel as a cleaning agent for shorelines contaminated by crude oil spills. Pure vegetable oil biodiesels (rapeseed or soybean) were shown to have a considerable capacity to dissolve crude oil. In a separate study in Texas, a commercial biosolvent, CytoSol, based on vegetable-oil methyl esters similar to biodiesel, was shown to be effective in coagulating the crude oil and allowing it to float to the surface of the water, where it can be collected. CytoSol was licensed by the California Department of Fish and Game as a shoreline cleaning agent in 1997.[11]

Heating with Biodiesel

In the United States, until fairly recently, biodiesel has been promoted mainly as a fuel for diesel-powered vehicles. But many people (even in the biodiesel community) don't realize that biodiesel can also be used as a heating fuel additive or replacement in a standard oil-fired furnace or boiler. That's because Number 2 heating oil (another middle-distillate fuel) is virtually the same as standard petrodiesel vehicle fuel, and biodiesel can be mixed at any percentage with Number 2 oil. When used for space heating, biodiesel is sometimes referred to as biofuel or bioheat in the United States. The conversion process for an oil-fired furnace or boiler is just as simple as the conversion for a diesel engine—just add the biodiesel to the fuel tank. No new heating appliance or expensive retrofitting is required. What's more, using biodiesel as a home heating fuel has virtually none of the cold-weather operating problems that are associated with using biodiesel in vehicles during the winter. Using biodiesel as a heating fuel is such a simple idea, you have to wonder why nobody thought of it sooner. Actually, someone did, because biodiesel has been used as a heating fuel in Italy, France, and a number of other European countries for many years. Now the United States is beginning to catch up.

Special Considerations

As mentioned previously, biodiesel has high solvent properties and tends to dissolve the sludge that often coats the insides of old fuel tanks and fuel lines. When used in a heating system this can potentially cause a clogged fuel filter or burner head, so biodiesel should be added carefully at first to old heating systems. Until all the sludge in the fuel tank has been dissolved, keeping an extra fuel filter on hand might also be a good idea for the first heating season.

Biodiesel should be stored in an indoor (or underground) storage tank because biodiesel, like Number 2 heating oil, will gel if stored outside in extremely cold weather. The pour point (the temperature below which the fuel will not pour) must be kept in mind if biodiesel is used. The pour point for Number 2 fuel oil is -11°F (-24°C). Although the actual pour-point temperature for biodiesel varies, depending on its concentration and original feedstock, it is consistently higher than that of Number 2 fuel oil. Consequently, biodiesel fuel should be stored at temperatures above its pour point.

Some people have actually used B100 to heat their homes with no problem. However, other people burning high percentages of biodiesel have experienced seal failures in their fuel pumps. The leaky rubber seal (or pump) can usually be repaired or replaced by a heating service technician in a short time, but the potential for this problem should be kept in mind if a high concentration of biodiesel is used. Some oil burner manufacturers are testing new seal materials to eliminate this problem in future burner models.

Great Potential

The potential for reduced reliance on imported Middle Eastern oil with the increased use of biodiesel as a heating fuel additive is substantial. In fact, if everyone in the northeastern United States used just a B5 blend, it could save 50 million gallons of regular heating oil a year, according to officials at the USDA Agricultural Experiment Station in Beltsville, Maryland. The experiment station has been heating its many buildings successfully with a biodiesel blend since 1999. At first the station staff used a B5 blend, but in 2001, encouraged by the test results, they switched to B20 and haven't experienced any problems. "Using biodiesel offers an opportunity to reduce emissions, especially particulate matter and hydrocarbons, and that's a great advantage," says John Van de Vaarst, deputy area director, who is responsible for facilities management and operations. "I used to refer to biodiesel as an alternative fuel, but now I call it an 'American fuel, made by American farmers.' I think it's an obvious strategy to help clean up the environment and reduce our dependency on foreign oil."[12]

Another series of tests on the use of biodiesel for space heating was conducted at the Brookhaven National Laboratory on Long Island. Sponsored by the National Renewable Energy Laboratory and the U.S. Department of Energy, the 2001 test report found that biodiesel blends at or below B30 can replace fuel oil with no noticeable changes in performance. Burning of the blends also reduced emissions of carbon monoxide and nitrogen oxide.

"There has been a lot of interest, particularly in the Northeast, in using biodiesel as a home heating oil," says Jenna Higgins, director of communications for the National Biodiesel Board in Jefferson City, Missouri. "I think it's definitely a very strong potential market in the future."[13] Roughly three out of four homes in this country that use oil heat are located in the Northeast, so the potential for expanding the use of biodiesel in the region is substantial. But can biodiesel meet the increased demand for the heating market? Residential consumption of Number 2 heating oil in 2002 was around 6 billion gallons nationwide, according to the Energy Information Agency. Assuming that every homeowner in the United States currently heating with oil switched to B20 (admittedly a large assumption), that would require about 1.2 billion gallons of biodiesel. U.S. production of biodiesel could cover this amount in a few years, assuming that this was the only use for biodiesel, which, of course, is not the case. Still, bioheat is an excellent strategy to clean up fairly dirty emissions quickly and easily without large retrofit costs.

Real-World Tests

But is bioheat really safe and effective? Bob Cerio, energy manager for the Warwick, Rhode Island, school district says it is. The district has been running biodiesel fuel tests, originally sponsored by the National Renewable Energy Lab, in three of its schools since 2001. During the first heating season, the district ran three different percentages of biodiesel (B10, B15, and B20) as well as a Number 2 fuel oil control in a fourth school. "It just worked very, very well for us," Cerio reports. "We had three different types of burners, three different types of boilers, and three different sizes, so we had an opportunity to test a wide spectrum of capacities. With the smaller boilers, we were able to get similar test data to what people would be experiencing in their homes," he adds.[14]

After a successful first season, Cerio switched to a B20 blend in the test schools for the 2002–2003 heating season without any problems, and he's no longer experimenting with any other concentrations. A number of tests for boiler efficiency and emissions have been conducted. "We're not seeing any change in efficiency, but we're seeing a reduction in sulfur dioxide, nitrous oxides, carbon monoxide, and carbon dioxide," he reports. "We've also discovered that our boilers are running much cleaner, so that saves us quite a lot of work cleaning them." Cerio is enthusiastic about the use of biodiesel as a heating fuel. "It's a very easy match for home heating, particularly if you have an indoor storage tank," he says. "Other than that, there really isn't anything that has to be done in order to use it."

Another biodiesel field trial involving about one hundred homes, sponsored by the U.S. Department of Energy, the New York State Energy Research and Development Authority (NYSERDA), and the National Oilheat Research Alliance, is being conducted by Abbott & Mills Inc., a fuel oil dealer in Newburgh, New York. Now in its third heating season, the tests with B20 are progressing well, according to Ralph Mills, the company's general manager. "So far, we have no news to report, which is good news," he says. "We've had no service problems associated with the fuel at all. The conclusion that we've come to at this point is that B20 is a viable replacement for traditional fuel oil."[15] Continuing the experiment to the next level, Abbott & Mills is now heating its office with B100. NYSERDA is expected to begin two more bioheat-related tests in the near future in other locations.

Clearly, bioheat works. But there are two main obstacles to heating a home in the United States with biodiesel: price and availability. Biodiesel generally costs more than Number 2 heating oil. How much more depends on who the supplier is and the quantity purchased. Nationally, the price of pure biodiesel ranges from about $1.60 to $3.00 per gallon, depending on the time of year and supply/demand. A B5 blend, however, should be only a few cents per gallon more than regular Number 2 heating oil.

Finding a local source of biodiesel for home heating can be a problem in many areas. There are about twenty major producers (and numerous small producers) of biodiesel scattered around the country as well as hundreds of local distributors, but the vast majority of the distributors are clustered in the Midwest, where biodiesel feedstocks are grown. Finding a biodiesel source in other parts of the country can be difficult. And locating a fuel oil dealer that offers biodiesel home deliveries has been a real challenge, even in New England, where 2.2 billion gallons of heating oil are consumed every winter. But that's beginning to change.

Catching On

The state of Maine is well known for its rugged winters, and bioheat would seem to be an obvious choice for the state's many environmentally conscious citizens. One industry pioneer capitalizing on this is Frontier Energy Inc. of South China, Maine. In 2002, sensing a new market opportunity, this offshoot of Frontier Oil Company began to offer biodiesel to homeowners in its regular delivery area between Augusta and Waterville. The company is currently offering—and actively promoting—a B5 "Basic Bioheat" blend as well as a B20 "Premium Bioheat" blend. For those who want it, B100 is also available, although the company doesn't recommend using it as a heating fuel at that concentration.

"It's going very well so far," says Joel Glatz, Frontier Energy's vice president. "We're probably selling about the same amount for vehicular use as we are for heating use at this point, but I think the heating application is what is really going to catch on in this state. We use about 300 million gallons in Maine for heating oil and about 150 million gallons for transportation annually, so, obviously, there is a much larger market for heating in this state."[16] Frontier's bioheat marketing strategy has worked; homeowner response has been extremely positive. "It's been fantastic," Glatz reports. "Those who have used it love it. The comment I usually get is, 'I can't tell the difference,' which is exactly what you want to hear." In early 2004 Frontier Energy reached an agreement with the C. N. Brown Company of South Paris, Maine, to provide the state of Maine with 40,000 gallons of B20 to heat state office buildings.

In Vermont, also famous for frigid winters and many environmentally aware citizens, the bioheat industry is just beginning to warm up. A number of new businesses, in cooperation with the Vermont Biofuels Association, the Vermont Sustainable Jobs Fund, and other key players, have been working hard to help facilitate the delivery of biodiesel home heating blends in the near future.

Vermont's Alternative Energy Corporation (VAEC) of Williston, Vermont, is another new venture in the Green Mountain State. Launched in early 2003, the company was awarded a grant from the U.S. Department of Agriculture for the "Steps Towards a Biorefinery Industry in Vermont" project. VAEC was also involved in a pilot heating project with the state of Vermont early in 2004.

In Massachusetts, bioheat is just beginning to show up on the radar screen. In the fall of 2003, Alliance Energy Services of Holyoke, Massachusetts, began to offer biodiesel as a heating fuel in a B20 blend. "Biofuel is readily available, and it makes sense for a lot of people," says Stephan Chase, the company's president. Alliance, which has been actively promoting its biofuel, has a growing number of customers. "It will be interesting to see what happens," Chase says. "The biofuel is a good product, and the Pioneer Valley has a lot of residents who are concerned about the environment, so it's a good combination; we should do very well with it here."[17] In 2004, Alliance began adding B3 to *all* its home heating oil at no additional cost to its customers.

Further south on the East Coast, where winter weather is somewhat milder, Tevis Oil of Westminster, Maryland, delivered its first B20 blend of heating oil to a residential customer in Upperco, Maryland, in November 2003. "This was the first delivery of soy biodiesel for use in home heating that we know of in this area," said Jack Tevis, president of S. H. Tevis and Son, which operates Tevis Oil. "Soy and other biodiesel fuels are used in the Midwest and New England to heat homes and run farm equipment, but it's still a fairly new concept here in the Mid-Atlantic region."[18]

The future of biodiesel as a home heating fuel looks good. It can provide farmers with a steady cash crop, help boost the economy, reduce dependence on foreign oil, and benefit the environment all at the same time. What's more, the use of bioheat is far beyond the experimentation phase. Expect to see many more homes heated with biodiesel in the near future.

PART TWO

Biodiesel *around the* World

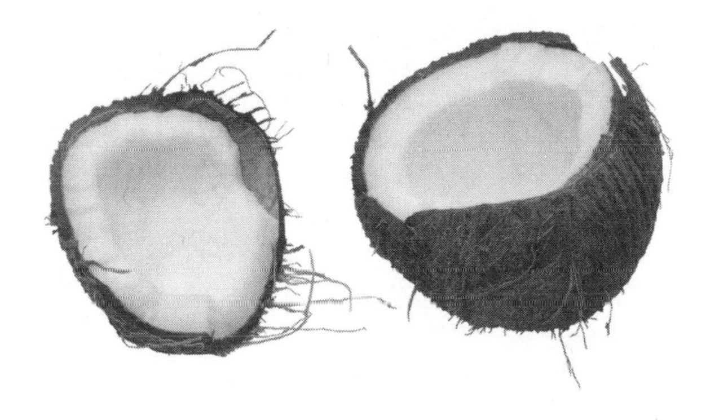

5

Europe, the Global Leader

Europe is the global leader in biodiesel production. Germany, France, and Italy combined produce nearly eighteen times more biodiesel than the entire United States. Biodiesel has been manufactured on an industrial scale in Europe since 1992, mainly in response to the supportive actions of various European Union (EU) institutions. Today, there are more than fifty plants in western Europe with a production capacity of about 2.2 million metric tons (627 million gallons) annually. These facilities are located primarily in Germany, France, Italy, Austria, and Spain, according to the European Biodiesel Board (EBB). Germany accounts for about half of the total EU capacity, with slightly over 1 million metric tons. France is second, with 502,000 tons; Italy third with 419,000 tons; Austria fourth with 100,000 tons; and Spain fifth with 70,000 tons capacity. *Actual* biodiesel production is roughly half of these figures; Germany produced about 650,000 tons, France about 366,000, Italy about 210,000, and Austria roughly 25,000 tons, according to EBB estimates for 2003. (Spain, a very recent entrant in the biodiesel market, did not have commercial-scale production estimates for 2003.) In eastern Europe, there are more than thirty biodiesel manufacturing plants of varying sizes; the Czech Republic leads with three large plants and thirteen smaller facilities, followed by Slovakia with one large plant and about nine relatively small facilities.

After the initial successes of the first biodiesel pilot plant at the Silberberg Agricultural College in Styria, Austria, in 1985, some small-scale commercial facilities began biodiesel production in other locations, but the quality

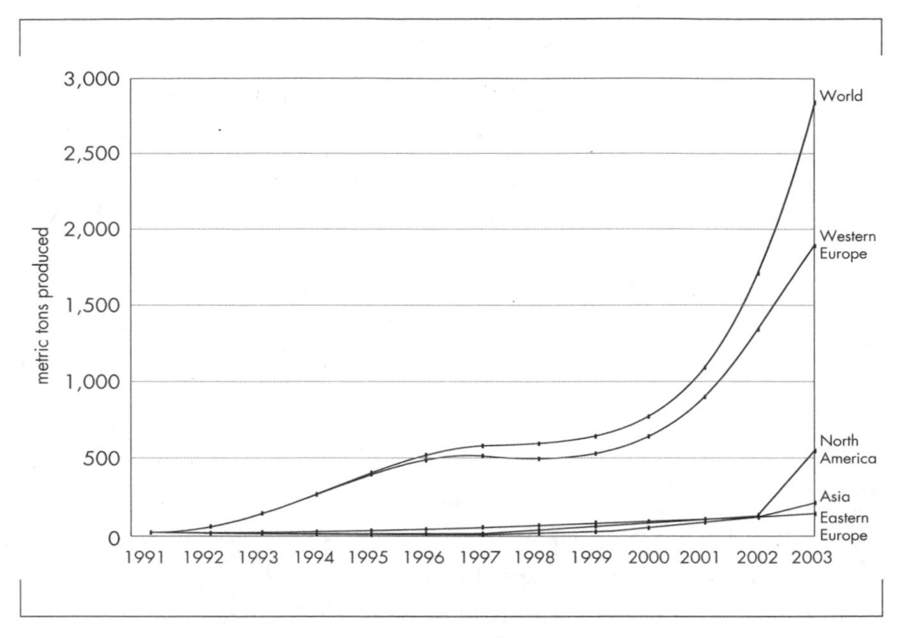

Figure 3. World Biodiesel Production (from the Austrian Biofuels Institute)

of their output was uneven. Other early plants capable of more consistent quality followed, and in 1991 the first industrial-scale biodiesel plant with a production capacity of 10,000 metric tons per year was started up in Aschach, Austria, along with some small-scale plants in France and the former Czechoslovakia. This was followed in 1992 by a much larger plant in Livorno, Italy, with an annual capacity of 80,000 metric tons.

The Common Agricultural Policy

In 1992 a number of changes made to the European Union's Common Agricultural Policy (CAP) had a dramatic impact on the biodiesel industry. In response to increasing surpluses of grains and other crops, the new policy established a set-aside program that prohibited farmers from growing food or feed crops on 10 percent of their arable land in the EU, while simultaneously allowing them to grow crops such as rapeseed, sunflowers, or soybeans on the set-aside lands for "industrial purposes." The production of biodiesel feedstock was one possible option under this provision, and beginning the following year biodiesel production began to grow dramatically. In 1992 Germany produced only 5,000 metric tons of biodiesel, while France produced virtually none. By the next year, German production had doubled to 10,000 metric tons, while French production had soared to 20,000 metric tons.[1] As demand grew, additional plants were constructed all over Europe in locations such as Leer, Germany (80,000 tons per year) in 1994, and Rouen, France (120,000 metric tons per year) in 1996, as well as in Sweden and elsewhere.

The new, expanded EU-25,[2] including the so-called accession countries (mostly in central and eastern Europe), will potentially increase the set-aside lands from the current levels of about 7 million hectares (17.3 million acres) to about 12 million hectares (29.6 million acres), offering the possibility for a large increase in biofuels feedstock production. Many of these new EU nations will undoubtedly be interested in using some of their agricultural lands for the production of renewable energy crops such as rapeseed or sunflowers as a way of limiting overproduction of food crops, especially grains, while still maintaining some income from what would otherwise be fallow land.

The European nations have not, however, relied entirely on virgin vegetable oil crops for all of their biodiesel feedstock. Used frying oils (UFO)[3] have played an increasingly important role in the picture. Although there were a number of limited experiments in prior years, the start of industrial biodiesel production from 100 percent UFO began in Mureck, Austria, in 1994. Subsequently, the UFO sector of the biodiesel market made especially impressive gains between 1998 and 1999. This was a time when set-aside lands for industrial crop production had been cut back to 5 percent, oilseed costs were high, and petrodiesel prices were at record lows, causing substantial losses for many biodiesel producers and prompting them to search for lower-cost feedstocks. They found what they were looking for with UFO. Although it is limited in supply, UFO does offer a relatively low-cost feedstock option for many producers.

But beginning in 1999 and continuing into mid-2002, vegetable-oil prices dropped again while petrodiesel prices were, on average, relatively high, causing a boom in the growth of biodiesel manufacturing plants, especially in Germany, where the number of plants increased by a remarkable 200 percent. Numerous construction projects were approved, especially in 2000, dramatically increasing the country's production capacity. Biodiesel production in Germany catapulted from 130,000 metric tons in 1999 to around 500,000 metric tons in 2002, with much of that increase taking place in the former East Germany. But by 2004, with vegetable-oil prices on the international market again moving higher, German biodiesel production capacity outstripped demand, causing considerable concern in the industry. "The current high vegetable-oil prices in Europe are definitely causing problems for biodiesel producers," says Raffaello Garofalo, secretary-general of the European Biodiesel Board. "This is happening at the same time that we have new legislation in the member states promoting biofuel. With the right foot we are going forward, but with the left foot we are lagging behind, and the gap between the two is getting wider and wider. It's an awkward situation."[4]

Production Issues

Since feedstock generally represents about 70 percent[5] of the direct production cost of biodiesel, it's clear that anything that affects the cost of vegetable oils will have a significant impact on the price of biodiesel made from these oils. The price of vegetable oil is driven by international markets beyond the control of biodiesel producers, and the inevitable ups and downs in those markets have caused the biodiesel industry significant challenges during periods of high vegetable-oil prices, like those that occurred in 1997–1998 and again in 2004. But in August 2000, for the first time in history, the crude petroleum oil price climbed above the price for vegetable oil on the world market. This phenomenon will likely be repeated more frequently as the world supply of petroleum oil continues to decrease and demand continues to increase. But regardless of the fluctuations in the international commodities markets, the historical disparity between the cost of production for biodiesel and that for petrodiesel has been substantial, putting biodiesel at a significant price disadvantage. For example, the cost of producing biodiesel in Europe was around 500 euros per 1,000 liters in 2001, while the cost of petroleum-based diesel was 200 to 250 euros per 1,000 liters.[6] This general price disparity is expected to continue for some time, and it is the main rationale behind various tax strategies that try to level the playing field. More on tax strategies in a moment.

There has been one development, however, that has offered biodiesel producers a little more flexibility in recent years. Initially, most of the biodiesel research in Europe was focused on rapeseed methyl ester (RME) and, later, on used frying oils (UFO). But after a lot of additional research and screening of potential feedstocks, it was determined that good biodiesel could be made from a wide range of feedstocks and even from multifeedstock blends (MFB). This discovery was one of the key factors in the subsequent development of modern biodiesel production facilities that can take advantage of the lowest-cost feedstocks or blends of feedstocks. Since the price and availability of various feedstocks can change rapidly (the ups and downs in the international vegetable-oil market in recent years are a prime example), designing manufacturing facilities that can alter their ingredients mix quickly has been extremely important. This design goal has been achieved in some of the most up-to-date facilities, which can select the desired "recipe" from a range of options stored in their process-control system. "The selection of an efficient process technology is very important," says Werner Körbitz, chairman of the Austrian Biofuels Institute. "Fortunately, there are many on the market right now, so one can choose."[7]

The other key factor in profitable biodiesel plant operation is yield. In most early plants the yield from the transesterification process was between 85 and 95 percent. While this may sound fairly impressive, it's important to understand that a drop in yield of only 10 percent can reduce profits by about 25 percent. Modern biodiesel plants convert all the free fatty acids and triglycerides in the vegetable oil and achieve a 100 percent yield.[8]

In a related issue, since the transesterification process produces roughly 10 percent glycerin as a by-product, the value of this substance also plays a role in the profitability of any biodiesel plant. The current world demand for glycerin is around 950,000 tons per year in a niche market that is sensitive to oversupply. Although this market has been able to absorb the approximately 100,000 tons of increased production resulting from recent growth in EU biodiesel production, continued increases could trigger a price collapse for this commodity. Trying to figure out what to do with the thousands of tons of excess glycerin from the continuing expansion of the biodiesel industry around the world will undoubtedly keep quite a few people awake at night for the foreseeable future.

The European Biodiesel Board

In 1997, as a sign of continuing development within the industry, the European Biodiesel Board (EBB) was formed. A nonprofit organization with the aim of promoting the use of biodiesel in the European Union, the EBB also serves as an organizational structure and collective voice for its member-producers. Located in Brussels, Belgium, the EBB has achieved a high degree of visibility and earned the confidence of EU institutions as well as nongovernmental organizations through its many activities. The organization has come a long way since its founding. "In the early 1990s, biodiesel was really an outsider in the fuel industry and was not well known," says Raffaello Garofalo, secretary-general of the EBB. "But today, with about 1.4 million tons of annual production, the biodiesel industry is well established in the EU, and we now have the involvement of big multinational corporations. That's a major achievement."[9]

Regulation and Legislation

The regulatory and legal frameworks covering biodiesel in Europe have developed gradually since the 1980s but have followed different paths in the various nations (even after the founding of the European Union). Austria developed the first RME biodiesel standard (ON C 1190) in

1991, followed by others in France and Italy in 1993, the Czech Republic in 1994, and Germany, with the most elaborate standard of all (DIN E 51606), in 1997. These standards were crucial in obtaining warranties for biodiesel use from numerous diesel vehicle manufacturers in these countries, resulting in increased demand for biodiesel as a vehicle fuel.

However, an even larger stimulus for biodiesel use was the Kyoto Protocol. Signed in 1997, the protocol committed the European Union to reducing greenhouse gas emissions by 8 percent from 1990 levels by a series of target dates from 2008 to 2012. But it wasn't until May 14, 2003, that the European Commission adopted the Directive for the Promotion of Biofuels (2003/30), which calls for the increase of biofuels' market share to 2 percent by 2005 and 5.75 percent by 2010. This directive was motivated by the need to reduce greenhouse gas emissions in the transport sector in response to the Kyoto Protocol, and to enhance energy security by reducing European dependence on imported oil. One significant provision of the directive (among many others) encourages greater use of biofuels in public transport and taxi fleets.

Overall, the directive should give the increased production of biodiesel feedstocks a significant boost, especially in the accession countries in the recently expanded EU-25. In most of the EU these goals are viewed as both realistic and achievable, but not without considerable effort and commitment. The French Agency for Environment and Energy Management (ADEME) has estimated that reaching the 2010 objective would require an increase in industrial rapeseed plantings from their current 3 million hectares (7.4 million acres) in the EU to 8 million hectares (19.8 million acres). Since the directives involved specify only sales rather than production, if local supply within any nation cannot meet supply it is possible to import the difference.[10]

In addition to the Directive for the Promotion of Biofuels, another main force driving increased biodiesel use in the EU is the Directive on Fuel Quality. The EU Directive on Fuel Quality (and a number of voluntary agreements under other programs) has resulted in significant advances in diesel engine technology that have improved energy efficiency and reduced emissions. These improvements require high standards for the fuel used by these engines, and the new EU CEN (European Committee on Standardization) fuel standard EN 14214, developed in cooperation with automotive, oil, and biodiesel industries in the member nations, ensures biodiesel's continued consistent high quality. First proposed in the late 1990s, the standard finally came into effect in late 2003. The new standard is being used as a basis for the replacement of the various individual national standards mentioned earlier. "The new standard is a major achievement," says Raffaello Garofalo of the EBB. "It's a reference point for our entire industry."[11]

Since biodiesel has consistently been more expensive to produce than petrodiesel, national fuel taxation policy has also played a key role in the development of the industry in Europe. But up until fairly recently, these decisions have been made on a nation-by-nation basis. Germany, France, Italy, Sweden, Austria, and the Czech Republic have all had some form of biodiesel tax exemption. These same countries as well as Denmark, Finland, the Netherlands, and Norway have also had various types of "carbon taxes" on the books that have been designed to gradually reduce the use of fossil fuels and encourage the use of renewables. But the patchwork of individual laws created a bureaucratic nightmare as well as substantial differences in taxation levels among the various nations. Trying to resolve this tangled mess with a common European standard has been a longstanding goal.

This seemingly impossible task was finally realized with the EU Directive on Energy Taxation (2003/96). Since the late 1990s, the EU had tried to develop an EU-wide energy tax strategy for its member states. After years of wrangling and numerous attempts—especially on the part of the United Kingdom—to block the initiative, the European Council of Ministers finally unanimously adopted a new directive on energy taxation on October 27, 2003. The directive establishes, among other things, rules for the detaxation of biodiesel and biofuels. Article 16 of the directive enables member states to apply an exemption or a reduced rate of excise duty to all biofuels (in pure form or in blends) sold in the EU for a period of six years, beginning in 2004. In addition, the directive will make it much easier for individual countries to amend their fuel tax laws.

The directive was also designed to reduce the distortions in competition that existed between energy products. Previously only petroleum-based oils had been subject to EU tax legislation, and not coal, natural gas, or electricity; these latter forms of energy are now included. The directive was also intended to increase the incentive to use energy more efficiently and to allow member states to offer companies tax incentives in exchange for specific initiatives to reduce emissions.

"In practice, what the directive says is those member states of the European Union who intend to detax biodiesel can do so without asking for prior authorization from Brussels," explains Raffaello Garofalo. "This is very important because, until the directive was published and adopted, the member states had to pass through a very long procedure in Brussels in order to have biodiesel and biofuels detaxation." This procedure required the approval of all the other states, lengthy debates, and inevitable delays, according to Garofalo.[12] The new directive eliminates all of that—at the EU level, at least. In January 2004, in response to concerns raised by some of the new accession countries, the European Commission proposed a series of transitional measures to help these nations gradually come into harmonization with the new EU-wide tax directives.[13]

Germany, the European Leader

While Europe is the global leader in biodiesel production, Germany is currently the undisputed highest-producing nation in Europe (and the world), with an annual production of 650,000 tons (185 million gallons). At first Germany's biodiesel was intended for use primarily in German agricultural machinery, especially farm tractors, in a closed-loop cycle. In 1982 some early tests were conducted on farm-tractor diesel engines at the Institut für Biosystemtechnik (Institute for Agricultural Engineering) in Braunschweig. But the uses for biodiesel soon multiplied well beyond the agricultural community, and today biodiesel is used in Germany to power a wide range of vehicles as well as forestry machinery, various kinds of boats, and other types of equipment.

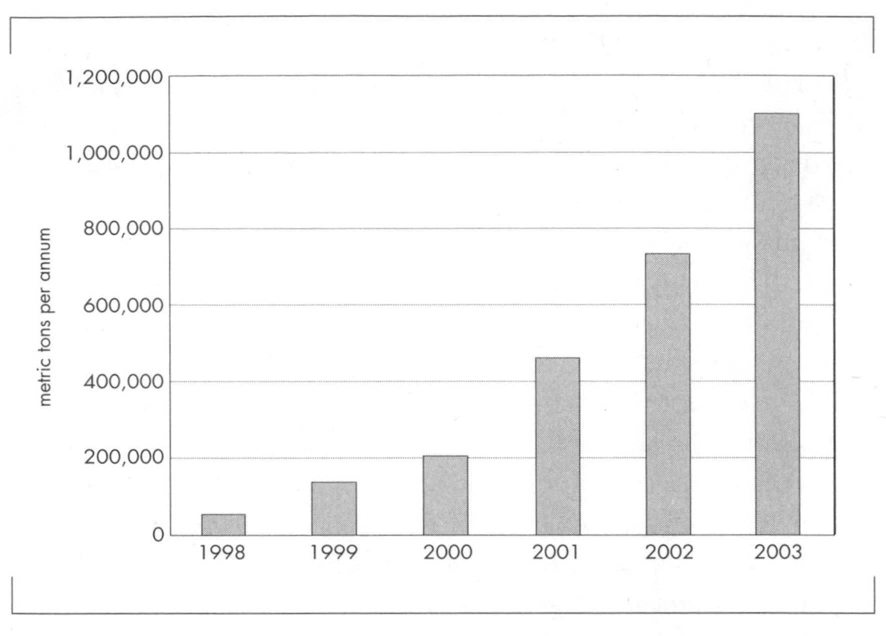

Figure 4. Biodiesel Production Capacity in Germany, 1998–2003 (from the Union zur Förderung von Oel- und Proteinpflanzen, or UFOP)

The first supplies of rapeseed methyl ester for early German trials were manufactured by the Henkel Company in Düsseldorf in a nondedicated chemical plant. A pilot plant was then established at the Connemann oil mill in the port city of Leer, located on the North Sea coast (Oelmuhle Leer Connemann GmbH & Co. has been in the seed oil business since 1750), and in June 1991 it produced its first ten tons of biodiesel. Three years later the pilot plant was replaced by an 80,000-ton-capacity (later increased to 100,000 tons), industrial-scale facility at the same location that used a so-called continuous deglycerolization (CD) process. This facility is now owned by the U.S. multinational Archer Daniels Midland (ADM). The following year a small-scale, 2,000-ton plant was opened by a farmer's cooperative in Grossfriesen in the former East German state of Saxony, and a short time later a 5,000-ton plant owned by a farmer's cooperative in Henningsleben in Thuringia began production.[14] In 1999 a 50,000-ton-capacity plant was started up at Ochsenfurt in Bavaria by Campa Biodiesel GmbH, and another facility, owned by Bio-Diesel

Wittenberge GmbH, began operating in Wittenberg in the former East German state of Mecklenburg-Vorpommern (Mecklenburg-West Pomerania), with a production capacity of 60,000 tons per year (since September 2000 this facility has been a joint venture between Bio-Diesel Wittenberge GmbH and Cargill Inc.).

The Building Boom

The post-1999 boom in German biodiesel plant construction mentioned earlier resulted in the addition of about twenty new production facilities across the nation. Oelmuhle Hamburg, a 100,000-ton-capacity plant owned by ADM, was built in Hamburg in 2001 based on the same CD production process used in ADM's Leer facility. Also in 2001, a 100,000-ton plant operated by Mitteldeutsche Umesterungswerke (MUW) GmbH opened in Bitterfeld; a 4,000-ton facility operated by BKK Bio-Diesel started up in Rudolstadt; a 40,000-ton plant owned by SARIA Bio-Industries GmbH began production in Malchin; and a 5,000-ton facility owned by PPM Umwelttechnik GmbH & Co. came online in Oranienburg.

In 2002 the construction boom continued with the opening of a 5,000-ton plant operated by BioWerk Sohland in Sohland; a 10,000-ton plant by Biodiesel Bokel GmbH in Bokel; a 35,000-ton plant by Petrotec GmbH in Südlohn; a 45,000-ton facility by Thüringer Methylesterwerke GmbH & Co. in Harth-Pöllnitz; a 100,000-ton plant by Natur Energie West in Marl; a 100,000-ton plant by Rheinische Bioester in Neuss; and a 100,000-ton plant by Nevest AG in Schwarzheide.

The building binge persisted in 2003 when Bio-Ölwerk Magdeburg opened a 50,000-ton facility in Magdeburg; EOP Elbe Oel started a 30,000-ton plant in Falkenhagen; Biodiesel Kyritz opened a 30,000-ton plant in Kyritz; Kartoffelverwertungsgesellschaft Cordes & Stoltenburg GmbH opened a 10,000-ton facility in Schleswig; Delitzscher Rapsöl GmbH & Co. opened a 5,000-ton plant in Wiedemar; and BioWerk Kleisthöhe GmbH started up a 5,000-ton plant in Uckerland.

In 2004 the rate of the expansion finally began to slow as German production capacity far outpaced demand. Nevertheless, a huge 130,000- to

150,000-ton-capacity plant belonging to Marina Biodiesel GmbH & Co. was under construction in Brunsbüttel, and a 37,000-ton plant owned by Rapsveredelung Vorpommern GmbH & Co. was completed in Malchin in May.[15] As of mid-2004, Germany has a total of twenty-five biodiesel manufacturing plants.

Pure Success

Biodiesel has been marketed in Germany mainly as a vehicle fuel (for both cars and trucks) at a "pure" B100 concentration (this policy was changed in late 2003 to include various blends of biodiesel as well). This strategy has been successful from a price standpoint because biodiesel has been exempt from fuel taxes, making it less expensive than petrodiesel for drivers of diesel cars and other vehicles (the 2003 approval of biodiesel blending has altered the tax reduction to correspond with the amount of biodiesel in the blend). Unlike France and Italy, however, Germany did not establish any upper limit on the amount of biodiesel produced that was eligible for tax exemption, which partly explains the huge expansion of production capacity in recent years.

Biodiesel production was given a boost in 1996 when the sale of leaded gasoline was banned in Germany. As a result, about one thousand fuel pumps at filling stations suddenly became available, and within a few months about six hundred of them had been converted to biodiesel use. In 1999 the German government introduced an additional ecological tax (eco-tax) on fossil fuels that increased in a series of steps over the following four years, making biodiesel even more attractive as an alternative fuel. The idea was to further reduce greenhouse gases while transferring the costs to polluters.

One of the earliest major successes for biodiesel took place when the German Taxi Association adopted the use of biodiesel nationwide after a successful 1991–1992 test in the city of Freiburg. The taxi association's practice of ordering large numbers of new taxis at one time helped spur Mercedes-Benz, Volkswagen, Audi, and Volvo to offer warranties for the use of biodiesel in their vehicles. The early collaboration of biodiesel producers with German automakers Mercedes-Benz, Volkswagen, BMW, and

MAN has been a significant factor in the success of biodiesel in Germany.

A number of public transport fleets in German cities have also switched to biodiesel. One of the best examples has been in the city of Heinsberg (located near the German border with the Netherlands), where the city works department, Kreiswerke Heinsberg GmbH, began its first tests with biodiesel in 1996. Over the next few years, the tests continued while the city also considered converting its buses to run on compressed natural gas. After all the pros and cons were carefully weighed, biodiesel won the contest. By 1998 the entire fleet of 130 buses (Mercedes-Benz and MAN) had been switched over to biodiesel, making Kreiswerke Heinsberg the first local public transport enterprise in Germany to operate its entire fleet on biodiesel.

The city of Neuwied, located beside the Rhine River, was the first municipality in the Rhineland-Pfalz region to convert its entire bus fleet to biodiesel. But unlike similar biodiesel projects in other cities, the Neuwied conversion involved the transformation of a mostly older fleet, in this case twenty-six 1982–1983 buses. While the age of the vehicles required that various seals and hoses on the diesel engines be replaced in order to make them serviceable for use with biodiesel, it still proved to be a cost-effective program. During the course of the project, which ran from January 1997 to October 2000, a number of the older buses were replaced with new models. One of the main criteria of selection for the new buses was that they be suitable for use with biodiesel. By the end of the study it was clear that the buses worked well on biodiesel, and the project was judged to be a success.[16] Other German cities are experimenting with biodiesel in their municipal fleets as well. Since the middle of 2003, the city of Wuppertal (famous for its monorail) has been testing biodiesel in fourteen of its buses. Würzburg, located in Bavaria, has been involved in a biodiesel bus trial as well.

But biodiesel isn't being used just on dry land. Since the early 1990s some tourist boats have been using B100 as their fuel on the Bodensee (Lake Constance, located along the German-Swiss border), Europe's second-largest drinking-water supply. The yacht clubs and individual boating enthusiasts using the lake have also embraced biodiesel and have included its use as part of their efforts to keep the lake clean and free from toxic fuel spills. A three-year pleasure boat study initiated in 2000 was conducted on the Bodensee with twenty-four participants. The study found only minimal problems related to softening of rubber hoses and gaskets on some older-model diesel engines. It concluded with a recommendation for further marine use of biodiesel on the lake.[17]

Support Organizations

Biodiesel has also benefited from promotional efforts by organizations such as the Union for the Promotion of Oil and Protein Plants (UFOP) that have emphasized biodiesel's locally grown feedstocks, environmental advantages, and high quality standards. As a result, biodiesel is widely associated with the brilliant yellow rapeseed fields that produce the feedstock. The UFOP's "Germany's most beautiful oil fields" publicity has helped propel biodiesel's popularity, and the fuel now accounts for about 3 percent of the German diesel market. There are currently more than 1,800 fueling stations offering biodiesel across the country, and the number keeps growing. It is estimated that there is potential for about one thousand additional biodiesel outlets in Germany. About 40 percent of the nation's biodiesel is sold at these fueling stations, while the remaining 60 percent is marketed through fleet operations for public transportation and commercial freight haulers.[18]

Germany's first preliminary fuel standard for biodiesel made from plant oil methyl ester (PME), DIN (Deutsches Institut für Normung) V 51.606, was published in 1994 and included a wider range of possible oil feedstocks than the earlier 1991 Austrian standard. The first German standard was followed by a second, improved standard (DIN E 51.606) for fatty acid methyl ester (FAME) in 1997. The importance of maintaining high quality standards for biodiesel production was clearly demonstrated around 1999. Due to the rapid expansion of both production and demand, biodiesel of inferior quality began to enter the market, causing engine problems and complaints from both consumers and auto manufacturers. In response, with the assistance of the UFOP, biodiesel producers created an organization called the Association for Quality Management of Biodiesel

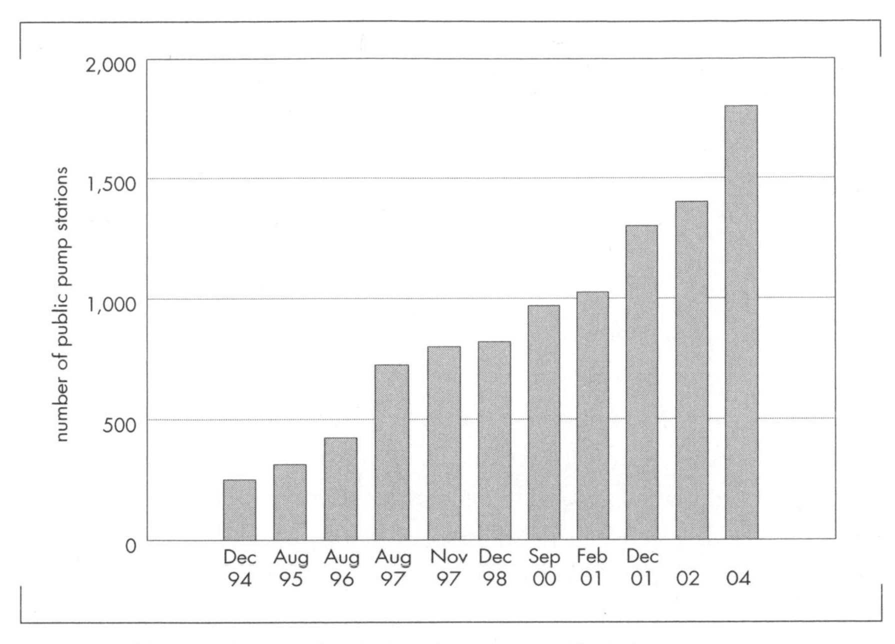

Figure 5. Public Pump Stations for Biodiesel in Germany, 1994–2004 (from the Union zur Förderung von Oel- und Proteinpflanzen, or UFOP)

(Arbeitsgemeinschaft Qualitatsmanagement Biodiesel, or AGQM) to ensure that only high-quality fuel would be sold. The group developed a quality seal that is displayed on fuel pumps containing approved biodiesel.

There are many other organizations that support the German biodiesel industry as well, but they are simply too numerous to mention.

Crop Issues

As in most of Europe, rapeseed is the main feedstock for biodiesel in Germany. Around 1974–1975, the first varieties of so-called 0-rapeseed were introduced, followed in 1986–1987 by improved 00-rapeseed (canola) varieties, which turned out, more or less by accident, to have very favorable characteristics for biodiesel. Initially the oleochemical industry consumed most of the annual German rapeseed crop, but after 1999 the biodiesel industry became the most important processor. This trend is expected to accelerate as demand for biodiesel grows. German rapeseed breeders continue to work on additional improvements to enhance certain genetic traits by breeding for new oil properties and increased resistance to disease; they also hope to increase the oil yield to 3.5 to 4 tons of oil per hectare (1.4 to 1.6 tons per acre). The best reported yield for rapeseed of 2.9 tons of oil per hectare was reported in 1999 by a farmer in Schleswig-Holstein. Improvements in precision plowing, soil preparation, fertilizer application, pest control, and harvesting will make the most of the genetic advances.

The former East German states account for about 60 percent of the annual German rapeseed harvest. In terms of the production of nonfood crops, rapeseed is the largest crop with 340,000 hectares (840,174 acres) devoted to it, while sunflower is a distant second with 5,300 hectares (130,968 acres). The UFOP estimates that in the near future, using available agricultural lands, it would be possible to reach 1 million hectares of rapeseed production, resulting in about 1.2 million tons of biodiesel.[19]

Straight Vegetable Oil

In recent years there has been some use of virgin rapeseed oil as a diesel fuel substitute in Germany. The oil either is used in vehicles with modified engines or is mixed with petrodiesel in the fuel tank. Engine manufacturers have serious doubts about this practice, however, and virtually all of the big players in the biodiesel industry take a dim view of this strategy. "The position of the EBB is very clear on this," Raffaello Garofalo says. "One barrel of bad biodiesel is enough to spoil the reputation of the entire biodiesel industry. The auto industry doesn't want to build special engines, which means that straight vegetable oil cannot be an option for the main market, except for those individuals who want to convert their engines. If you use straight vegetable oil in a normal diesel engine it will work for about six months if you are lucky, but then the engine will be ruined. We have had big problems with this in Germany. Some people were putting straight vegetable oil in their cars, and then after the engine seized up, they went to the manufacturer claiming that biodiesel had caused the problem. This was having a tremendous negative impact on the biodiesel industry."[20] (See "Straight Veg" in chapter 10 for more on this issue.)

Despite these and other growing pains, the biodiesel industry in Germany is generally headed in the right direction. Although German biodiesel production capacity has grown exponentially in recent years, once demand catches up it is likely that the industry will continue to expand on a somewhat slower but more sustainable basis. The long-term economic, environmental, agricultural, and security advantages of biodiesel production in Germany should ensure that the industry has a bright future.

France

After Germany, France is the second-largest producer of biodiesel in the world, with an annual production of approximately 366,000 metric tons (104 million gallons). (Between 1993 and 1999 France was the leading European producer of biodiesel, but it was overtaken by Germany in 2000.) Two main biofuels have been promoted in France to meet biofuels targets set by the European Commission for transportation fuels: ethanol and biodiesel.[21] Taken together, these two biofuels represent about 1 percent of total fuel consumption in France. France produces roughly three times more biodiesel than it does ethanol. This is mainly in response to the fact that nearly 70 percent of cars and trucks in France are now diesel powered. France's strategy for implementing greater biodiesel use has been to blend it with petrodiesel fuel and heating oil, usually at a B5 concentration. Consequently, unlike their counterparts in Germany and Austria, where biodiesel has been promoted and sold mainly as a differentiated product, many biodiesel users in France may not even be aware of the presence of biodiesel in their fuels. Biodiesel is also sometimes blended at rates between B5 and B30 for use in French vehicle fleets. Approximately four thousand vehicles in various communities are involved in municipal fleet programs across the country. Public buses in Paris, Bordeaux, Dijon, Dunkirk, Grenoble, and Strasbourg are now using a biodiesel blend.

Early Activity

In 1981 feasibility studies for the use of fatty acid methyl esters (FAME) as fuel in diesel engines and heating boilers were conducted in France by the French Petroleum Institute (IFP) and the Agency for Environment and Energy Management (ADEME). At about the same time, PROLEA, one of the major French agricultural associations, began a study that focused on the diversification of outlets for agricultural products. Oilseed crops were included in the study. Although these early French efforts eventually resulted in some of the earliest biodiesel legislation in Europe, the use of biofuels in general, and biodiesel in particular, did not start to develop on a commercial basis until after the European Union set-aside rules for agricultural lands were implemented in 1992. In the same year, concerns about the potential for pollution caused by large-scale farming of rapeseed in France helped spur the adoption of the so-called Agro-Environmental Charter.

A major series of advanced laboratory and field tests was conducted in the early 1990s on the use of rapeseed methyl ester (RME) in diesel engines by the ADEME and IFP. Most biodiesel in France is produced from RME, although sunflower ester has been undergoing testing and evaluation since 1995. In France, biodiesel is generally referred to by the name *diester*, a contraction of *diesel* and *ester*. (*Diester* is also the trademarked name of the rapeseed methyl ester produced by Diester Industrie, the nation's major biodiesel supplier.)

The first biodiesel in France was produced in an existing methyl ester plant in Péronne, owned by Castrol, and another in Boussens, owned by Sidobre-Sinova. Following the success of a small-scale pilot plant in Compiègne, the first dedicated biodiesel plant was built in the same location by the Robbe Company in 1993 (the plant is now owned by Robbe/Diester). In 1995 the production capacity of the plant located in Boussens (now co-owned by Cognis/Diester) was enlarged. Additional expansion of the French industry occurred in 1995, when Italian-based Novaol added 70,000 tons of capacity to an adapted chemical plant in Verdun. That was followed by a 120,000-ton-capacity dedicated biodiesel plant (the largest in the world at the time) in Rouen, a joint venture between Sofiproteol and VaMo Mills.[22]

Government Policies

Virtually all biodiesel in France is made from rapeseed oil grown specifically for that purpose, mainly on set-aside land. The rapeseed pulp produced after the oil has been extracted is used for animal feed. As in the rest of Europe, the EU's Common Agricultural Policy with its set-aside lands provisions has had a major impact on French biodiesel production. The reformed EU Common Agricultural Policy adopted in June 2003 sets a "carbon credit" payment of 45 euros per hectare for growing nonfood crops. But French oilseed farmers feel this is not enough support for them to grow industrial crops and believe that 90 to 100 euros per hectare would be necessary to maintain stability in nonfood production in their country. The debate continues.

Beginning in 1993, the French government has supported biodiesel by offering full excise-tax exemption, but for limited state-set quotas (this strategy has been followed in Italy as well). These quotas, however, have acted as a damper on the expansion of biodiesel production capacity. The tax exemptions on biofuels in France sparked a legal challenge in 2000 from the multinational oil company BP, which claimed that the subsidies were damaging its business. The European Court of Justice initially sided with BP. But the French government appealed the decision and petitioned the European Commission for approval of tax breaks on biodiesel used for heating and transport fuel. The commission ultimately approved the request, using a different clause of a 1992 mineral oils (petroleum) directive.[23] As of 2003, the quota for tax-exempt biodiesel was 317,500 metric tons. However, there was an overproduction of 47,500 tons that did not qualify for the exemption, and this amount was either exported or used in the chemical industry for other purposes. Since 2000, French biodiesel producers have been waiting for a 70,000-ton increase in the quota.[24]

The Main Players

The top biodiesel producer in France is Diester Industrie, which controls about 80 percent of the French biodiesel industry. (Novaol represents the other 20 percent.) Diester owns the largest European biodiesel plant, which is located in Grand-Couronne, Normandy, and now has an annual production capacity of 250,000 metric tons (70 million gallons), of which about 180,500 tons are approved for the excise tax exemption. To keep these figures in perspective, this one French facility has about three times the capacity of the annual biodiesel output for the entire United States. The plant, which cost 18 million euros, was built in 1995 with participation from the French oilseed industry (Sofiproteol), grower cooperatives, and local authorities. Diester Industrie also owns the expanded biodiesel plant in Compiègne, which has 60,000 tons of approved production. In the past few years, production at these two plants has increased significantly.[25]

The popularity of diesel vehicles in France has created a somewhat unusual situation for French petroleum oil refiners, who now suffer from surpluses of gasoline and shortages of diesel fuel. The surplus gasoline must be exported (primarily to the United States) at low prices, while petrodiesel must be imported, sometimes at higher cost. This helps explain why the French refiners are generally happy to blend as much biodiesel into their petrodiesel as the French government and EU regulations will allow, since it reduces the need for expensive imports.[26] Some large oil companies like TotalFinaElf are among the biggest consumers of biodiesel, and virtually all French biodiesel is sold to petroleum companies for blending purposes.

In 1994 the Club des Villes Diester (Club of Biodiesel Cities), an educational and collaborative network of biodiesel users, producers, and institutions, was formed to help promote greater use of biodiesel. Members include oil companies, auto manufacturers, professional and trade associations, and national nonprofits, as well as thirty communities across the nation. The group recognizes that, since the 1970s, French industry has made significant progress in reducing emissions, leaving the transport sector as a continuing major source of urban air pollution. The French Clean Air Act (Loi sur l'Air) of 1996 requires reductions in urban pollution. The Club des Villes Diester has supported a wide range of biodiesel initiatives (especially in public transport) to help achieve the goals set out in the law.

The arrival of McDonald's fast-food restaurants in 1979 was viewed by many French citizens as a foreign invasion (bordering on gastronomic sacrilege). Today, however, these same restaurants have begun to play a productive role in the biodiesel industry. In July 2003 McDonald's France, along with its partners Ecogras and Sud Récupération, which collect used frying oils from McDonald's, signed an agreement with the Italian biodiesel producer Novaol for the reprocessing of McDonald's frying oils into biodiesel. Ecogras and Sud Récupération are supplying Novaol with approximately 1,200 tons a year of oils collected in French McDonald's restaurants, which use partially hydrogenated rapeseed oil. The biodiesel made from the used frying oil is exported to Italy, where it is used as a heating oil.[27]

The French strategy of blending biodiesel seamlessly into the diesel fuel sector has been a success and has created a mutually beneficial partnership between biodiesel producers and the traditional petroleum industry. The French biodiesel industry is well established and has matured over the years to the point where it now has carved out an important niche in the nation's energy portfolio. The industry's proven track record should allow it to continue to grow to help meet the biofuels targets set by the EU for 2005, 2010, and beyond.

6

Other European Countries

Although Germany and France are the two largest players in the European biodiesel industry, a number of other nations, such as Italy, Austria, the Czech Republic, and Slovakia, have also played significant roles in the development and growth of the industry, although they have frequently followed slightly different paths. Here are some of the highlights of those developments.

Italy

Italy ranks third in world biodiesel production with about 210,000 tons (60 million gallons) produced annually. Italy is highly dependent on energy imports, which amount to about 80 percent of total energy consumption. This makes Italy vulnerable in the energy sector, but it also offers biodiesel producers considerable long-term prospects for continued growth. At the moment, however, liquid biofuels (mostly biodiesel) represent less than 0.5 percent of Italian renewable energy sources, which in turn represent only about 9 percent of the nation's total energy picture. In the past decade, Italy has promoted the increased use of all types of bioenergy, including biodiesel, and that emphasis is expected to grow as the nation attempts to meet the goals of the Kyoto Protocol and the EU.

The earliest biodiesel production in Italy started with some small, existing, nondedicated methyl-ester plants in the northern part of the country. In 1992, two Italian companies, Novaol (then named Novamont) and Estereco, took part in a joint venture to develop a biodiesel industry, funded in part by the European Commission. A large-scale, dedicated plant with a capacity of 80,000 tons per year was built by Novaol in the port city of Livorno.[1] The Livorno facility, which processed both rapeseed and sunflower oil, was one of the first large-scale commercial biodiesel plants in the world. Initially the entire output of the plant was directed to the heating oil market, but some of it was subsequently marketed as transport fuel. Unlike the case in most other European nations, in Italy both transport fuels *and* heating oil are taxed at fairly high levels. Since heating oil provides biodiesel producers the same tax advantages as vehicle fuel, but at slightly lower quality levels, most Italian producers have opted to focus on the heating oil market.

Biodiesel blends sold for use in Italy have been exempted from fuel taxes, but only up to certain production quotas established annually by the government. In prior years the limit had been set at 125,000 metric tons per year, but in 2001 the limit was raised to 300,000 tons. In addition, Italy established a "carbon tax" in 1999 on CO_2 emissions, which increased the excise tax on fossil fuels in a series of steps through January 2005. Two-thirds of the funds derived from the tax are managed by regional authorities to finance energy efficiency and renewable energy initiatives and other ecological projects, while the other third is administered by the national government. Although most of the funds have been used to promote a wide variety of other renewable energy initiatives such as photovoltaics, the net result has been to make alternatives to fossil fuels such as biodiesel more attractive. The EU's new Directive on Energy Taxation (2003/96) will undoubtedly affect some aspects of the Italian carbon tax (as well as similar tax laws in other EU member nations).

In early 2000 the Italian government passed a biofuels initiative, the Biomass Fuels National Program (Probio), which began to implement the resolutions contained in the Kyoto Protocol. In cooperation with regional authorities, the program allocated about 2.5 billion euros over three years to help stimulate the production of biofuels from agricultural feedstocks. In 2000 about 80 percent of Italy's biodiesel was produced from (mainly imported) rapeseed oil, while about 20 percent was from sunflower oil, most of which was domestically produced. In April 2001 a National Voluntary Agreement for the use of biofuel in national transport was signed. The agreement anticipates the introduction of up to 5 percent biodiesel in Italian public transport.[2]

There are currently nine producers of biodiesel in Italy. Novaol in Livorno has received the largest share of production quotas in recent years, followed closely by Fox Petroli in Ancona and then by Bakelite Italy in Solbiate; Comlube in Brescia; Consorzio Agr.E.E. in Città di Castello; Oleifici Italiani in Bari; Distillerie Palma in Naples; IGS in Milan; and Defilu in Milan. In July 2003 Novaol announced plans to double the capacity of its Livorno plant with the installation of a new continuous production line. The project, representing an investment of 7 million euros, was completed in 2004.

Novaol, a member of the Cereol Group, is one of the leading biofuels enterprises in Europe. With headquarters in Milan, Novaol also has biodiesel production facilities in France and Austria. In 2000 Novaol produced and sold more than 200,000 tons of biodiesel, roughly translating into the production of 200,000 hectares of oilseeds (rape or sunflower). In Italy Novaol's biodiesel is marketed under the brand name Diesel-Bi.

The vast majority (about 90 percent) of biodiesel produced in Italy is used for heating purposes. A smaller share of biodiesel is used for transport fuel at a B5 blend and for some fleets at a B30 blend, particularly for use in public transport. Public transit buses in Florence, Gorgonzola, Padua, and Perugia now run on biodiesel. As a heating fuel in Italy, biodiesel is generally used in its pure B100 form (but sometimes as B20), and even the Vatican is reported to have biodiesel-fueled boilers. About half of Italy's total biodiesel production is sold to petroleum oil companies for blending, while the rest is marketed through other channels. Continued growth in the Italian biodiesel industry seems likely, especially in the vehicle fuel sector.

Austria

Austria is the fourth-largest EU producer of biodiesel, with about 25,000 tons (7 million gallons) produced annually. Despite the country's relatively modest production figures, it would be hard to overemphasize the role that Austria has played in the biodiesel industry. The first research and development projects in the biofuels sector in Austria began as early as 1973. As noted in chapter 2, much of the most important and sustained early biodiesel research was conducted at the Austrian Federal Institute of Agricultural Engineering (BLT) in Wieselburg and at the University of Graz in the early 1980s. As an outcome of that research, one of the first industrial-scale biodiesel plants in the world—with a production capacity of 10,000 metric tons per year—was started up in Aschach, in the state of Upper Austria, in 1991. "Unfortunately, it was also the first one which was closed because of oil prices that were lower than $10 per barrel at the time," notes Manfred Wörgetter of BLT.[3]

One of the key private-sector collaborators with some of this early work was Vogel & Noot GmbH of Graz, which, beginning around 1982, was active in developing technologies to produce biodiesel from plant and animal oils and fats. In 1988 the company received a patent for a transesterification process that used waste cooking oils and waste fats for the production of biodiesel. In 1991 Vogel & Noot provided the rapeseed oil process technology for a 1,000-ton biodiesel plant in Mureck, and the following year the company was involved in the development of a larger-scale biodiesel facility in Bruck with a capacity of 15,000 tons (more recently expanded to 25,000 tons), now operated by Novaol Austria. Vogel & Noot were also connected with the construction of a number of small farmers' cooperative biodiesel plants in various Austrian communities that process different types and qualities of oils and fats.

In 1994 the plant in Mureck was altered and expanded by Vogel & Noot to enable the facility to produce biodiesel from used frying oil collected from the surrounding region. The biodiesel output from the multi-feedstock plant is now used for high-quality vehicle fuel. In 1996 the biodiesel division of Vogel & Noot was sold and a new company,

BioDiesel International (BDI) Anlagenbau GmbH, was created. BDI, with headquarters in Graz, has subsequently been involved in a number of multifeedstock plant projects in Germany, Spain, the United Kingdom, and the United States.

A new biodiesel plant located in Zistersdorf, in the state of Lower Austria, was opened in April 2002 by Biodiesel Raffinerie GmbH, a subsidiary of DonauWind KEG. The technology for the continuous transesterification reactor process of the 40,000-ton, multifeedstock facility was supplied by ENERGEA Umwelttechnologie GmbH (also a DonauWind subsidiary) of Klosterneuburg, Austria. DonauWind, which began operations in 1997, has a four-turbine wind farm in Zistersdorf as well as another turbine near Vienna. Several additional wind farms in other locations are planned. DonauWind is a good example of an increasing number of renewable energy companies that are expanding the scope of their activities to include a wide range of renewables.

Another new biodiesel facility was opened by Bioenergy-Biodiesel Erzeugung GmbH in Wöllersdorf in 2002. The 20,000-ton-capacity plant relies on used frying oil as its main feedstock. Finally, the most recent addition to Austria's biodiesel industry was in Arnoldstein, where Biodiesel Kärnten GmbH opened a 25,000-ton, multifeedstock plant in 2003. All together, there are nine commercial biodiesel plants in Austria, with their production capacities ranging from 1,500 to 40,000 tons per year. The majority of Austrian biodiesel is produced from rapeseed oil, with sunflower oil, used frying oil, and animal fats playing lesser roles.

Early on, Austria adopted a 95 percent tax reduction for biodiesel, as long as the fuel was used full-strength (B100) in vehicles. As of January 1, 2000, all fuels from renewable sources are free of petroleum fuel taxes. Biodiesel has been widely marketed in Austria to large municipal bus fleets, transport and taxi companies, and other diesel users, especially in environmentally sensitive locations. What's more, there are currently more than one hundred filling stations in Austria that offer biodiesel to the general public.

One of the earliest and most successful biodiesel experiments in an Austrian city transit fleet took place in Graz. Beginning in 1994, the

Grazer Verkehrsbetriebe (GVB) began field tests with two of its city buses running on biodiesel made from recycled frying oil. The three-year test was considered a success, and in 1997 eight additional buses were switched to biodiesel, followed in 1999 by ten more. By 2002 GVB had 55 of its 140 buses running on biodiesel. The plan is to convert the entire fleet to biodiesel by 2005.[4]

In 1991 the Austrian Standardization Institute developed and implemented a biodiesel standard (ON C 1190) for rapeseed methyl ester (RME), becoming the first country to do so, and in 1997 the institute published a more sophisticated standard (ON C 1191) for fatty acid methyl ester (FAME). The former helped encourage biodiesel warranties for diesel tractor engines from numerous manufacturers, while the latter opened the way for use of a wider range of biodiesel feedstocks, including used frying oil. The emphasis in the latter standard was in defining the quality of the fuel that ends up in the fuel tank rather than focusing on the feedstock from which it was made. One of the indirect results of that second standard is that about 1,400 tons of used frying oil is collected every year from 135 McDonald's restaurants in Austria and then turned into biodiesel for the city buses in Graz.

Although the biodiesel industry in Austria is fairly mature, there is still room for additional growth.

Czech Republic

Until fairly recently the Czech Republic had the largest number of biodiesel production plants in the world, totaling sixteen, with a combined production of about 50,000 tons annually. However, with the exception of the 30,000-ton Milo plant in Olomouc, the 12,000-ton Bionafta plant in Mydlovary, and a 3,000-ton Agropodnik facility in Jihlava, the plants were small-scale farmers' cooperatives with less than 1,000 tons capacity. The developing Czechoslovakian (and later Czech) biodiesel industry was highly influenced by the early research and development work in Austria, and consequently the Czechs followed the

Austrian model of small farmers' cooperatives. The large-scale plant in Olomouc was completed in 1994 with the technical assistance of the Austrian biodiesel plant designer Vogel & Noot.

Initially biodiesel was used in the Czech Republic at full strength (B100) as vehicle fuel, but later on a variety of blends—especially B30—also became popular. In 1994 the Czech standard (CSN 65 6507), developed with the assistance of Austria, was published for rape methyl esters. A subsequent standard for biodiesel blends in the B30 to B36 range was published in 1998. Engine warranties for biodiesel have been issued on a wide range of Czech cars, trucks, and tractors.

Biodiesel generally enjoys full detaxation in the Czech Republic. Motorists can refuel their biodiesel-powered vehicles at over 176 fueling stations across the country. In January 2001 a new Czech regulation (86/2001) was instituted to try to harmonize various government agricultural support programs with EU practices regarding nonfood agricultural production and set-aside lands.[5] With the acceptance of the new European standard EN 14214 in 2004, the previous Czech standards were replaced. Based on a solid decade of experience, the Czech biodiesel industry should continue to grow as the EU-25 nations attempt to meet the new biofuels targets in the years to come.

Slovakia

Slovakia has also had a long history of biodiesel activity. Like the Czech Republic, Slovakia has a large number of smaller, mostly farmers' co-op plants (500 to 1,500 metric tons per year) and a few larger-scale facilities. The first small-scale plant in Slovakia was built by a group of technicians from the ZTS-Martin company (formerly known as a manufacturer of heavy weapons and tanks) in collaboration with the University of Bratislava in 1992. Most of the small-scale plants that followed were supplied with process equipment from the Slovak company MTD.[6]

Today, the larger Slovak producers are Palma-Tumys a.s. in Bratislava (60,000 tons), AgroDiesel in Revúca (1,500 tons), and BIO BHMG in

Spišský Hrušov (1,500 tons). There are also about six smaller plants with production capacities of about 500 tons scattered throughout the country. Total annual biodiesel production capacity in Slovakia is around 75,000 tons.[7] Palma-Tumys in Bratislava has been an industrial manufacturer of fats and vegetable oils since the 1920s, and the company's biodiesel plant now uses the same CD production process found in ADM's facilities in Leer and Hamburg, Germany.

The Slovakian biodiesel standard (STN 656530), which went into effect in 2001, should be replaced by the new EU standard, EN 14214. Biodiesel has been used both as a vehicle and heating fuel in Slovakia. B30, sometimes referred to by the name *bio-naptha*, is the most popular blend of biodiesel in the country. Until fairly recently biodiesel enjoyed full detaxation in Slovakia, but that changed in 2002 when a new national excise tax law came into effect that caused considerable confusion and removed some of the previous incentive for the use of biodiesel blends, resulting in a drop in demand.[8] The structural changes in the Slovak economy in recent years and the transition to EU membership have proved challenging and often difficult for the country. However, if these transitional problems can be overcome, the long-term outlook for the Slovak biodiesel industry is reasonably good.

Spain

While many European countries have been developing their biodiesel industries for over a decade, Spain has been slow to jump on the bandwagon. Just a few years ago Spain didn't even show up on European biodiesel statistical charts. This is partly due to the fact that, until fairly recently, Spain has focused much of its attention on bioethanol, and as a result it is now the largest EU producer of that fuel. But in the past few years, in response to the new EU targets for biofuels percentages, Spain has been making up for lost time in the biodiesel sector. Spain's late entry into the biodiesel market may actually work to its advantage, since it can avoid many of the early challenges experienced by its pioneering neighbors and make use of the latest technological advances, especially multifeedstock processes. What this means is that much of Spain's biodiesel can be made from relatively inexpensive used frying oils, especially olive oil, which the country has in great abundance. Over 1 million tons of plant oils are consumed in Spain every year, about half of which end up as used frying oil.[9]

In 1999 the company Grupo Ecológico Nacional (GEN) in Barcelona participated in a pilot project (sponsored by the Austrian Biofuels Institute) to recycle waste fats under the EU's ALTENER-2 renewable energy initiative (managed by the European Commission's Directorate-General for Energy, ALTENER-2 is the second phase of this ongoing initiative). The goal of the project was to establish a low-cost system for recycling used frying oils from restaurants, hotels, and households and then using them as biodiesel feedstock.[10] Following up on that program, biodiesel was first introduced to the Spanish market for tests, mainly in pilot projects with public transport buses. In September 2002 biodiesel was tested in three buses in Navarre with the aim of encouraging future biodiesel use in additional buses and also in municipal garbage trucks.

In October 2002 Spain's first dedicated biodiesel production plant was opened by Stocks del Vallès in Montmeló. The 6,000-ton-capacity, multifeedstock plant, which cost 5 million euros to construct, utilizes used cooking oil as its primary feedstock and was designed so it could easily be expanded to 18,000 tons by 2005. That was followed in 2003 by a 20,000-ton-capacity plant in Álava built by Bionor at a cost of 7 million euros and a 50,000-ton-capacity plant in Reus constructed by Bionet Europa costing about 8 million euros. In February 2003 the first public biodiesel pump in Spain opened in the Catalan town of Tárrega. Located in a Petromiralles service station, the pump dispenses B30 manufactured by Stocks del Vallès. Petromiralles plans to install another five biodiesel pumps, one each in the towns of Figueres, Igualada, Vilafranca del Penedès, and Cercs and one in the port area of Barcelona.

Plans for additional Spanish biodiesel plants continue to pop up in numerous other locations. The Corporación Energía Hidroeléctrica de Navarra S.A. (EHN), with headquarters in Navarre, is scheduled to complete a 16-million-euro, 35,000-ton-capacity plant in Caparroso in late

2004. EHN is also planning to invest 30 million euros in a 100,000-ton-capacity plant in El Morell in the Catalan province of Tarragona, which will use rapeseed, soy, and sunflower oils as feedstock. The plant is expected to open in 2006. It's interesting to note that EHN is one of the main developers of wind power in the world and is also a player in the fields of biomass, hydroelectric power, and solar energy. The company has recently added biofuel production to its diverse portfolio. Companies like EHN (and DonauWind in Austria) clearly understand where the energy economy of the twenty-first century is heading.

In a cooperative venture between the Department of Chemical Engineering of the University Complutense of Madrid and the Institute for the Diversification and Saving of Energy (IDAE), a 4,000-ton-capacity biodiesel production plant that will depend on used frying oil and virgin oils as its main feedstocks is planned for Alcalá de Henares. The output of the plant is intended to be used in the city's municipal buses.[11] But that's not all. A huge 250,000-ton-capacity biodiesel plant is under construction by another consortium, AES Corporation/Sauter Group, in the Mediterranean port of Cartagena. The 50-million-euro facility is scheduled for completion by the middle of 2005. According to the IDAE's *Plan for the Promotion of Renewable Energy in Spain, 1999–2010*, the target for biodiesel production in Spain by 2010 is 400,000 tons per year, and the country appears to be well on its way to meeting or exceeding that goal.

Belgium

Belgium has had a long history of biodiesel involvement. But unlike many of its neighbors, it has not followed through with widespread national implementation because of lingering reservations about the environmental and economic performance of the fuel. A 1937 Belgian patent for the use of ethyl esters of palm oil and an experiment with a commercial bus operated between Brussels and Louvain in 1938 were two of the earliest biodiesel-related activities anywhere in Europe. In addition, the article "The Use of Vegetable Oils as Fuel in Engines" was published in Belgium in 1952.[12] More recently, between 1991 and 1992 an experiment was conducted in Mons with fifteen buses belonging to Transport En Commun (TEC)-Hainaut running on B20 (the most common biodiesel blend in Belgium). The drivers were satisfied with the performance of the buses, and the only negative comments received were from a few people who were bothered by the "barbecue" odor of the exhaust.[13]

In 1992 a biodiesel pilot plant was built by the De Smet Company in Feschaux, and for the next three years a number of additional tests with buses and other vehicles running on the fuel produced by the plant were conducted, with favorable results. In 1996 a two-year test was conducted by the Flemish Institute for Technological Research (VITO), in cooperation with the University of Graz in Austria, on used vegetable oil as a biodiesel feedstock. Despite all of this activity, the majority of Belgian biodiesel output (which amounted to 5 percent of the European total in 1998) was exported to France, Germany, and Italy. Today, Belgian biodiesel production has ceased.

The two main producers of biodiesel in Belgium were BASF Group in Feluy (ex Pantochim Sisas production site) and Oleon. BASF took over operation of the former Sisas production site complex in Feluy, which included a 60,000-ton-capacity biodiesel plant, in July 2001. But the following year BASF decided to discontinue biodiesel production at Feluy. Oleon has two biodiesel production plants, one in Olegem (60,000 tons) and the other in Ertvelde (40,000 tons), both previously owned by Atofina/Petrofina, but neither plant is currently producing biodiesel. Nevertheless, Belgium has a substantial biodiesel capacity that could be put back into production without major additional investment.

Denmark

Although Denmark was involved in some successful biodiesel tests with buses in Copenhagen in the mid-1990s, only limited progress has been made in the country's biodiesel sector due to a longstanding lack of political support for the use of biofuels in transportation. Nevertheless, in 2002 Denmark produced 10,000 tons of biodiesel and presently has a production capacity of about 30,000 tons at the Emmelev plant in Otterup. Denmark would seem to be well suited for future growth in biodiesel production and use, since both its climate and agricultural sector are ideal for high yields of rapeseed oil. What's more, most Danes are environmentally conscious and most likely would embrace greater use of biodiesel as well as recycling programs for used frying oil, which at present is an underutilized resource.

The Netherlands

ATEP Nederland b.v recently signed an agreement with BASF Nederland b.v to take over BASF's industrial site in Arnhem. ATEP intends to construct a biodiesel production facility with a capacity of 100,000 tons per year on the site. The plant, which will use the continuous transesterification process developed by Oelmühle Leer Connemann Gmbh & Co. in Germany, was supposed to be completed in 2004. However, the plans have been put on hold, and an alternate location is being studied. ATEP is also actively working to increase public awareness of biofuels and to boost the sale of biodiesel throughout the country. The company's sales emphasis will focus on water sports, construction equipment, and the farming sector, as well as fleet owners such as taxi and bus companies.[14] Biodiesel promoters in the Netherlands hope that the government will implement the new EU guidelines on biofuels and eliminate taxation on biodiesel.

Sweden

The southern part of Sweden offers excellent potential for the production of rapeseed, but for many years most of the emphasis on Swedish biofuels has been directed toward ethanol made from wheat and other biomass. Stockholm, in fact, has Europe's largest fleet of ethanol-fueled buses. As a result, the use of biodiesel in Sweden is not as widespread as in many other European countries. Nevertheless, in 1992 the first Swedish biodiesel production plant was opened by Svenska Ecobränsle AB in Klippan; it produces about 1,000 tons annually. Rapeseed is the primary feedstock, but some tests have been conducted with linseed, lard, and used frying oil. Around 2001, the plant's capacity was expanded to 8,000 tons annually.

Biodiesel is fully tax-exempt for a quota up to 30,000 tons. About 80 percent of the total Swedish biodiesel production is used as B100 for vehicle fuels, while most of the remainder is blended at a B2 level with petrodiesel. More than nineteen public biodiesel fuel pumps presently serve Swedish drivers. There are numerous diesel engine warranties for biodiesel use in Sweden, based on the Swedish biodiesel standard (SS 1554 36). The Swedish Farmer's Federation (LRF) supports increasing biodiesel production in the future for use in Swedish ferries due to the fuel's relatively low impact on marine environments.

United Kingdom

Despite the fact that the United Kingdom consumes 15 million tons (4.2 billion gallons) of petrodiesel fuels a year, it has been one of the slowest adopters of biodiesel in Europe. Perhaps it's the isolating influence of the English Channel. Some say it's politics. "British farmers are ready and willing to supply this new developing market and are extremely frustrated when they see biofuels promoted and encouraged elsewhere in the world," said National Farmers' Union spokesman Rad Thomas in December 2003. "We are in the absurd situation where we export oilseed rape to Germany for biodiesel production, while we are short on diesel supply in the U.K."[15] In all fairness it should be noted that in August 2002, a reduction of 20 pence per liter in the fuel tax was implemented by the government to help level the playing field for biodiesel producers in England. And in May 2004 the government launched a twelve-week consultation aimed at setting a minimum sales target for biofuel beginning in 2005 and deciding whether a "renewable transport fuels obligation" should be imposed on fuel producers.[16] Still, many in the agricultural community and biodiesel industry have complained for years that the government has had a "muddled and unfocused policy" regarding biodiesel.

Regardless of what may (or may not) be happening in the national government regarding biodiesel, quite a bit has been going on at the local level, where rapidly increasing numbers of diesel-powered cars offer a significant opportunity for biodiesel producers. UK Cargill has had a relatively small, 2,000-ton-capacity plant located in Liverpool for several years. In addition, a group of entrepreneurs have set up a number of small-scale biodiesel plants that rely mainly on used frying oils for their feedstock. These producers have developed a modest but growing customer base, including municipal governments, local truckers, supermarkets, and breweries, as well as individual drivers. Ironically, finding enough used frying oil—rather than customers—has been the main problem for these producers. But biodiesel activity has expanded beyond the small local producer. There are now more than sixty-seven biodiesel forecourt (filling station) locations in England that sell biodiesel blends to the general public. London-based Greenergy has begun marketing a B5 blend it calls GlobalDiesel in southwestern England and more recently at twenty Tesco forecourts in the southeastern part of the country. The Green Shop garage, located near Stroud, Gloucestershire, became the first garage in the United Kingdom to sell biodiesel in July 2002.[17]

There is also quite a lot of recent biodiesel activity in the northeast of England, where area farmers are being encouraged to create "onshore oil fields" by growing biodiesel feedstock crops to help boost jobs and income in their rural economy. Similar initiatives have been promoted in other rural parts of England as well. What's more, Petroplus, the owner of a crude oil refinery in North Tees, has been processing around 20,000 tons of biodiesel (called Bio-Plus) a month for use in commercial vehicles, and the plant could soon require up to 180,000 tons a year of oilseed rape to meet its expansion plans. Petroplus reportedly accounts for about 60 percent of the current United Kingdom biodiesel market share.[18] In April 2004

Greenergy unveiled plans for a new 100,000-ton-capacity, multifeedstock plant. The £10-million (15-million-euro) continuous-process facility will be constructed in partnership with Novaol at a yet-to-be-determined site, but most likely on the Humber estuary in the northeast region.[19]

But the really big news in the British biodiesel industry is Biofuels Corporation's huge 250,000-ton-capacity plant at Seal Sands, Middlesbrough, on the northeastern coast. The facility, described as the largest and most advanced in the world, will cost an estimated £25 million (37 million euros) and is scheduled for completion in early 2005. A second 250,000-ton-capacity unit is tentatively planned at the same site for completion later in 2005, which would bring the combined capacity for the facility to an impressive total of 500,000 tons annually—and would exceed the total biodiesel output of France. ENERGEA Umwelttechnologie GmbH of Klosterneuburg, Austria, is providing the continuous-process design engineering. The biodiesel produced at Seal Sands will be mixed at 5 percent with ultra-low-sulfur petrodiesel. Farmway, an independent farmers' cooperative, is part of a consortium of companies that is looking at building a vegetable-seed-crushing plant to supply the raw materials for the proposed Seal Sands plant.

Scotland

In Scotland, there was not much biodiesel activity until the United Kingdom biodiesel tax reduction of 2002. However, since then there have been some significant developments. In November 2002 Rix Petroleum introduced Rix Biodiesel (a B5 blend with petrodiesel) to the Scottish market. The fuel was an instant hit at the company's twenty-five Scottish outlets, and sales of the biodiesel blend now account for about 90 percent of the company's total diesel sales. The biodiesel blend is doing so well, in fact, that some of the Rix stations have eliminated standard diesel fuel entirely.

But the most significant activity in Scotland is taking place in Newarthill near the city of Motherwell in Lanarkshire, where a 50,000-ton-capacity biodiesel facility that will use tallow as its main feedstock is being built next to an existing animal-rendering plant by Argent Energy. BioDiesel International (BDI) of Austria will provide the overall planning and delivery of the multifeedstock esterification and glycerin-processing equipment for the £15 million (22.5-million-euro) facility, which is scheduled to open in 2005. After a number of delays, the plant construction began in March 2004. When completed, the facility will produce enough biodiesel to supply about 5 percent of Scotland's requirements for diesel.

Northern Ireland

In Ireland, there is growing interest in the use of waste frying oils and beef tallow as biodiesel feedstocks, especially since the BSE (mad cow disease) crisis and the resulting prohibition of using fats in animal feed enacted in January 2001. Recent studies have identified about 10,000 tons of used frying oils as well as 2,000 tons of rendered tallow that could be used as biodiesel feedstock. There are now three small-scale biodiesel plants in Northern Ireland: Biofuels Northern Ireland in Belfast; Molloy Fuel Services in Milford; and O'Neill's Fuels Ltd. in Coalisland. In addition, a number of small-scale vehicle-testing programs have been conducted in cooperation with the EU ALTENER program.[20]

Lithuania

Oilseed rape is the largest oil crop in Lithuania, with an area of 60,000 hectares (148,000 acres) given over to its cultivation in 2002 for domestic vegetable-oil production and export. However, only a few hundred tons of this crop was processed domestically for biodiesel. There is quite a lot of fallow land in Lithuania, some of which is well suited to the production of industrial energy crops.[21] Consequently, in early 2003 the Kaunas Technical University in Kaunas, Lithuania, began working on a biodiesel research project related to testing a variety of known and new oilseeds, with assistance from the EU's ALTENER-2 project.

Interest in biodiesel clearly extends beyond the laboratory. The Rapsoila company constructed a $4.6 million biofuels plant in the Mazeikiai region of Lithuania in 2002. The 100,000-ton-capacity rapeseed oil methyl ester facility came online in early 2003.[22] Approximately 1,500 small farms in the surrounding region are reportedly supplying the rapeseed for the plant.

Prospects for future growth of the biodiesel industry in Lithuania seem reasonably good.

Latvia

A new 10-million-euro biodiesel plant is reportedly under construction in the Latvian port of Riga. The 75,000-ton-capacity facility will be operated by Baltic BioDiesel Ltd. and will use primarily rapeseed oil as its feedstock. Construction of the refinery will be carried out with technical assistance from Germany's Campa-Biodiesel GmbH, and the plant is scheduled for completion in 2005. Most of the biodiesel produced will initially be sold to customers in other EU countries, with only 10 to 15 percent earmarked for Latvian consumption. The project is intended to help increase the demand for rapeseed and develop the local agricultural economy.[23] Biodiesel is also reportedly being produced in the town of Valka (in northern Latvia) by Delta Riga Ltd.[24] There seem to be reasonably good prospects for future expansion of the Latvian biodiesel industry.

Poland

The first research work and testing of rapeseed methyl ester in Poland was conducted between 1989 and 1991 at Wyzsza Szkola Inynierska (WSI–College of Engineering) in Radom. Preliminary road trials on a Tarpan car were also conducted. Between 1991 and 1994, this work was continued and expanded at the Institute of Aviation in Warsaw, where the

Austrian standard ON C 1191 was used as a reference for testing. In 1996 a small biodiesel pilot plant was built in Mochelek by the Instytut Budownictwa Mechanizacji I Elektryfikacji Rolnictwa (IBMER) Institute of Warsaw, financed by the Polish Committee for Scientific Research. The output from the Mochelek plant was mixed with petrodiesel to form a B5 blend distributed by CPN S.A. Wroclaw.[25]

In the third or fourth quarter of 2004, two new biodiesel plants are expected to start production: PKN Orlen's Rafineria Trzebinia, located in Trzebinia, and Mosso, a food-processing company with a new oilseed-pressing plant located near Warsaw. The Trzebinia plant opened in October. By the end of 2004 it is estimated that total Polish biodiesel output will reach about 6,000 to 7,000 tons.[26] Since Poland has generally been a major producer of rapeseed, there is good potential for further growth of the Polish biodiesel industry.

Bulgaria

Bulgaria has one commercial biodiesel plant, which is located in Brousartsi near the border with Romania and operated by Sampo Inc., headquartered in Sofia. Originally a vegetable-oil-processing facility, the 3,500-ton-capacity plant was converted to biodiesel production in 2001 with a simplified transesterification process and now relies primarily on used frying oil (UFO) as a feedstock. But local supplies of UFO have not been large enough to meet the needs of the plant, and the company has been trying to obtain additional supplies from neighboring countries and possibly even from the United States. The biodiesel produced by Sampo has been used mainly at B50 and B100 concentrations in trucks, buses, and automobiles in the region. Some of the glycerin by-product from the plant has been refined for use in pharmaceuticals, while some has been used as a liquid fertilizer and also as an ingredient in hydraulic fluid. During the winter months, when gelling is a problem, Sampo sells much of its biodiesel output to a Swedish company located in Bulgaria that manufactures briquettes and matches.

Sampo has spearheaded the formation of the National Biofuels and Renewable Energy Sources Association. Biodiesel production has begun to attract the attention and interest of various government agencies, and three other commercial groups are reportedly considering possible Bulgarian biodiesel projects in the future.[27]

More European Activity

Before moving on to the rest of the world, it should be noted that there has been some biodiesel activity in a number of other European countries.

- Switzerland has a small-scale biodiesel plant located in Etoy with a capacity of 2,000 tons. Swiss biodiesel is generally used as B100 in farm machinery.
- In Norway biodiesel is used for vehicle fuel, but most of it is imported. There are currently more than eighteen public biodiesel fuel pumps serving Norwegian drivers.
- Yugoslavia has had an on-again, off-again biodiesel production history related to trade embargoes. Yugoslav biodiesel production peaked in 1995 at 5,500 tons but then returned to zero in 1996 after the embargo was lifted.

There have also been some experiments, tests, feasibility studies, or actual road trials of biodiesel in Portugal, Greece, and Hungary, but to date there are no known large-scale production facilities in these countries. Small-scale activity, however, is undoubtedly taking place.

7

Non-European Countries

Although Europe has been the main focus of most biodiesel activity for many years, there have been a lot of interesting developments in many countries elsewhere as well. When the official survey of the international status of the industry was published in 1997 by the Austrian Biofuels Institute for the International Energy Agency, twenty-four countries were included. Since then, that number has increased to at least thirty-nine. Especially in the past few years, biodiesel activity has begun to expand exponentially as more and more governments, concerned about the growing volatility and increasing prices in the international oil market, discover the benefits of biodiesel. But in the absence of strong leadership on renewable energy issues at the national level in some of these countries, individuals or small groups of citizens as well as nongovernmental organizations have taken matters into their own hands to initiate biodiesel projects. However, because many of these grassroots activities have tended to take place "under the radar screen" of the international press, it is sometimes difficult to assess just how much of this activity is going on. The survey that follows is an attempt to fill in at least some of the larger blank spaces on the global biodiesel map. The journey begins with Africa and follows a meandering but generally eastward course toward the Pacific Ocean and the Americas.

South Africa

South Africa has had a long but inconsistent history of biodiesel activity. As mentioned in chapter 2, South Africa was involved in early biodiesel experiments prior to World War II, which were subsequently abandoned in favor of coal and synthetic fuels. A second round of research and testing took place between 1980 and 1984 and made some significant advances. But, again, the work was abandoned. In recent years biodiesel has made yet another comeback, though it still has a long way to go before it achieves widespread market acceptance, since South Africa is still heavily dependent on coal-based fuels.

A June 2001 report published jointly by the Council for Scientific and Industrial Research (CSIR) and the Agricultural Research Council investigated technologies that could provide alternatives to crude-oil-based fuels. Biofuels were highlighted in the report as offering a good deal of potential. A follow-up report published in March 2003 recommended placing a greater emphasis on biodiesel in South Africa. In addition to decreasing the country's dependency on imported fossil fuels, the report said that greater biodiesel production would strengthen the agricultural sector and create new jobs in rural areas, a key national priority.[1] The Department of Science and Technology has recently been coordinating the national effort to introduce a viable biofuels industry in South Africa.

The South African potential for biodiesel production is enormous, according to a recent United Nations study. With the use of 2.3 million hectares of land (5.7 million acres), it is estimated that 1.4 billion liters (517 million gallons) of biodiesel could be produced, without having an adverse impact on food supplies. Even half of this projected amount could supply about 17 percent of current road and rail use of petrodiesel in the country. The study suggested a more modest (but still ambitious) target of 10 percent within ten years.[2] Another study, commissioned by Earthlife Africa, Johannesburg, and the World Wide Fund for Nature, Denmark, noted that much of this production could come from a large number of relatively small biodiesel facilities spread across the country, offering significant employment opportunities for farmers and other local entrepreneurs.[3]

In order to supply enough biodiesel to create a B5 blend for the entire nation, 280 million liters (approximately 260,000 tons) of biodiesel would have to be produced in South Africa annually. Current production doesn't even begin to approach that figure. In Merrivale, located in the province of KwaZulu-Natal, Biodiesel SA, a company that was established in January 2001, produced up to 600 liters of biodiesel a day (about 100 tons per year) from used vegetable oil. The biodiesel produced in Merrivale is used mainly by companies that operate heavy earth-moving equipment, according to the plant's owner, Darryl Melrose. The plant has recently been upgraded to a capacity of 2,000 liters a day, but Melrose has not been able to find enough used cooking oil to meet all the demand for his biodiesel. As a consequence, Melrose has spent a good deal of time researching alternative feedstocks. He finally decided that the hardy *Jatropha curcas* (physic nut) would best suit his (and the industry's) needs. Melrose is now part of a private/public partnership to cultivate this prolific oilseed-bearing plant.[4]

"We're trying to empower poor farmers on the east coast of South Africa between Durban and Mozambique where the climate conditions are suitable," Melrose explains. "*Jatropha curcas* can really benefit the farmers we are working with because the seed crop is toxic, so it can't be stolen to eat; animals won't even feed on it because it's inedible. So, I think, given our particular situation in this country, it's the perfect crop for this purpose."[5] About 300 hectares are scheduled to be planted with jatropha. Field tests to date have been extremely positive, and jatropha is reportedly taking off like wildfire in various parts of KwaZulu-Natal. About one thousand growers are already involved, and their number is increasing.

Another, somewhat larger biodiesel production facility with a 5,000-liter-per-day (about 1,000 tons per year) capacity began operating in May 2002 in Wesselsbron, in the province of Free State. The plant, operated by sunflower farmer Johan Minnaar, makes use of locally grown sunflower seeds as well as used cooking oil. Minnaar was the first to produce a vegetable-oil-based fuel on a commercial basis in South Africa.

Even more significantly, the large oil and chemicals group Sasol is studying the feasibility of producing biodiesel in South Africa using soybeans as the main feedstock. The proposed biodiesel plant will most likely be located at Secunda in Mpumalanga and would produce about 80,000 tons annually.[6]

Various changes to South Africa's Customs and Excise Act have been proposed to allow the mixing of diesel and biodiesel to reduce the tax burden on the industry, and early in 2002 the minister of finance announced a 30 percent tax reduction on biodiesel. What's more, in 2003 the South African government proposed accelerated depreciation of 50 percent in the first year, 30 percent in the second year, and 20 percent in the third year of the cost of capital expenditures for biodiesel plant and machinery located on farms.

For the time being, South Africa has adopted the European standard for rape methyl ester. However, due to the different biodiversity and climate conditions in the country, a series of tests are being conducted on soy, sunflower, canola, groundnut (peanut), cotton, and jatropha in order to establish an appropriate South African standard. Interest in biodiesel in South Africa is growing, and the long-term prospects for the industry appear to be excellent.

Other African Countries

There has been interest in developing biodiesel initiatives in a number of other African countries where the cultivation of a variety of oil-bearing crops is seen as a way to strengthen the agricultural sector, generate jobs, and reduce expensive imports of petrodiesel. A number of studies have been conducted by universities and other organizations on various aspects of biodiesel production, but aside from a few small projects, this work has not been translated into any large-scale commercial biodiesel development (except for a $1.2 million plant under construction by Ghana Bioenergy Ltd. at Pomadze in the Central region of Ghana). In most cases, lack of funding has been a major obstacle.

However, lack of industrial-scale biodiesel activity does not necessarily imply failure, since biodiesel technology—unlike that of the petroleum industry—can be viable at almost any size or scale. Consequently, in many parts of Africa—especially rural areas—a different model makes a lot of sense. This alternative strategy involves local, community, or regionally based projects that rely less on sophisticated technology and more on hard work and determination. There is one project that has planted the seeds of this version of the industry—literally. Jatropha, moringa, and neem seeds are being used by the Africa Eco Foundation, a nonprofit organization based in South Africa, to implement a biodiesel-creating initiative in southern Africa. The seeds are used to create plantations, which can be harvested for oils and other biomass by-products. To date, around seventeen countries in Africa have established pilot programs, with the largest, in Ghana, composed of 220,000 hectares (543,620 acres). Mali has been involved in the program since 1964. As usual, the biggest problem facing the initiative has been lack of funding.[7]

In many African cities, charcoal is the most common fuel for low-income people. Charcoal production has been blamed for deforestation and serious degradation of forestland in a number of African countries. Not only could the adoption of biodiesel provide a renewable substitute fuel, but also it could simultaneously be used as a land reclamation strategy. The people of Africa could benefit enormously from the widespread development of local biodiesel initiatives that, in addition to the obvious environmental benefits, would offer more jobs and greater self-reliance.

India

India imports 70 percent of its oil (65 percent from the Persian Gulf), at an annual cost of $18 billion (813 billion Indian rupees). In the past few years, with those costs continuing to rise, the country has started to invest in biodiesel as a way of stabilizing energy prices and increasing its energy security. Since petrodiesel represents about 80 percent of Indian transport fuels, the potential market is enormous. But India is following a strategy uniquely different from that of many other nations. Much of the nation's future biodiesel industry will be powered by oilseed-bearing trees (OSBT), rather than the field crops preferred by the EU and the United States. And unlike countries that rely on palm oil, India will focus primarily on inedible oils. The country plans to use native plants such as jatropha, mahua, and karanji to produce the feedstock oil for its biodiesel. What's more, plans call for growing feedstock crops in many areas that are not presently in agricultural production. And in another departure from standard biodiesel wisdom, straight vegetable oils from OSBT will be used to fuel many rural electrical generators.[8]

The use of vegetable oil as a fuel is not a new idea for India. During the 1930s the British Institute of Standards in Calcutta examined a group of nonedible oils from indigenous plants as potential diesel fuels. In 1940 a textile mill in Warangal, in the state of Andhra Pradesh, powered its entire operation (and the surrounding community as well) with nonedible oils. These facts were generally forgotten until 1999, when the residents of the village of Kagganahalli, in the state of Karnataka, told Dr. Udupi Srinivasa, a mechanical engineering professor at the Indian Institute of Science and head of a rural development agency, about an inedible oil from the seeds of the honge tree (also known as the pongam tree or Indian beech) that their grandparents had used for lamp oil. Dr. Srinivasa immediately recognized the oil's potential as a biodiesel fuel. "Here we were—all scientists—looking at technical solutions like windmills, gasifiers, solar panels, and methane generators for rural India, and we had not made the obvious connection with the potential of nonedible oils known from Vedic times as fuels," Dr. Srinivasa said later.[9]

The rediscovery of this local biodiesel feedstock led one company, Dandeli Ferroalloys, to convert all five of its 1-megawatt diesel generators to run on honge oil, resulting in significant savings on their energy expenses. And thanks to a separate rural development project, the villagers of Kagganahalli now have water to irrigate their crops from pumps powered by honge-oil-fueled electrical generators. The previously dry and desolate village now produces watermelons, mulberry bushes, sugarcane, and grains.[10] The Kagganahalli project has been the catalyst for numerous other recent Indian biodiesel initiatives.

Deep in the tropical forests of southern India, the Kolam people of the isolated village of Kammeguda, in Andhra Pradesh, also have had their lives changed by biodiesel. Until recently the Kolam had no telephones, televisions, running water, or electricity. Then, in 2002, Dr. Srinivasa walked into the village and literally lit up their lives. He showed the Kolam how the seeds of the karanji trees in the nearby forest could be turned into biodiesel fuel to power an electric generator. The generator has brought electric lights as well as running water to the village. "With lights, we can chase away snakes and animals that stray into our village in the night. We can catch the occasional thief also," says Lakshmi Bai, who manages the tiny power station. "Earlier, we used to put our children to sleep early, but now we make them study under the lights."[11]

Back in the mainstream of national commerce, Indian Railways has also been experimenting with biodiesel. On December 31, 2002, the first successful trial run of a high-speed passenger train was conducted when the Delhi-Amritsar Shatabdi Express used a B5 blend as fuel. The test was successful, and India Railways plans to expand the use of biodiesel with higher-percentage blends in more locomotives on other lines in the near future. The feedstock for the biodiesel was the jatropha bush, which could be grown on either side of India Railways' tracks.[12] If this proposed strategy is implemented, India Railways trains will be running beside rows of the same plants that fuel its locomotives.

Indian biodiesel activity really began to heat up in 2003. In November, DaimlerChrysler announced a new public/private partnership in India for the production of biodiesel from jatropha on eroded ground. In addition to yielding fuel, the five-year plantation project is also expected to create new jobs and reduce CO_2 emissions. It is hoped that the plantations created by the test will subsequently be operated by municipal authorities and can serve as models for other parts of India as well as for other countries. Partners in the project include the University of Hohenheim (in Germany) and the Indian Central Salt & Marine Chemicals Research Institute. DaimlerChrysler is providing some of the money, test vehicles, and engineers for the project.[13]

On December 10, 2003, the Indian Oil Corporation announced that it would begin field trials of running buses belonging to Haryana Roadways on a B5 biodiesel blend with petrodiesel. About 450,000 liters of biodiesel will be used in the pilot project.[14] Public transit isn't the only sector that is moving toward biodiesel. Automobile manufacturers such as Mahindra & Mahindra Ltd. as well as Ashok Leyland have endorsed biodiesel as fuel for their vehicles. On December 31, 2003, a state (as opposed to national) government initiative for the construction of five biodiesel pilot plants was announced. Each plant project, which will cost about 16,850,000 rupees ($372,000), will be a partnership with a different regional organization. This initiative is in addition to a national government proposal to set up similar plants in cooperation with the Indian Oil Corporation.[15] India appears to be well on its way to establishing a widespread and diverse biodiesel industry that should benefit a broad spectrum of its citizens in many different ways.

Thailand

Since 2001, Thailand has had an active biodiesel program, officially inaugurated and sanctioned by the nation's king. Experiments have been conducted with various plant oils that could replace petrodiesel, such as palm, coconut, soybean, peanut, sesame, and castor. The government has empowered the Department of Trade Registration to set biodiesel standards for the entire nation, and the Excise Department has been studying tax exemptions for biodiesel products. For a number of years there had been numerous requests for national standards due to the uneven quality and different biodiesel formulations found in various parts of the country.

Biodiesel is used for many different purposes in Thailand, powering everything from cars, trucks, and small farm tractors to fishing boats and stationary engines. In May 2003 the Royal Thai Navy completed a series of biodiesel tests and reported that it was satisfied with the results. The navy intends to use biodiesel in some of its boats and vehicles in the near future. More extensive biodiesel testing is planned with palm oil and used cooking oil. Once the best formulation is determined, the biodiesel will initially be used in vehicles owned by the armed forces and the Energy

Ministry and later in city buses owned by the Bangkok Mass Transit Authority. Widespread commercial use of biodiesel, however, is not expected until 2006.[16]

China

The Liaoning Province Research Institute of Energy Resources in China reported trials with pure vegetable oil as well as transesterified oils between 1991 and 1994. The Chinese researchers focused on guang-pi, an oil-bearing tree that can be grown on marginal land.[17] In 1998 a pilot project partly financed by the European Commission was conducted in cooperation with the Austrian Biofuels Institute, the Center for Renewable Energy Development in Beijing, and the Scottish Agricultural College in Aberdeen, Scotland, as partners. The project evaluated the overall potential of feedstock available for possible future biodiesel production. The study looked at all known oilseeds, animal fats, and recycled cooking oils in China.[18]

In 1999 Hong Kong residents Keith Addison and Midori Hiraga were preparing to start on a two-year, 40,000-kilometer, overland journey through twenty-six countries from Hong Kong to Cape Town, South Africa, in a pioneering expedition in two biodiesel-fueled Land Rovers. After a good deal of publicity generated by their venture, the Hong Kong government began investigating the use of biodiesel in the city. But there has been stubborn opposition from some local officials, and the initiative has not progressed beyond a few tests.[19] However, as of 2003 several companies in mainland China are reportedly producing large quantities of biodiesel from rapeseed oil grown there. China's rapidly growing economy is contributing to the strong demand for oil and the continuing upward pressure on international oil prices. Because of the nation's huge population, if biodiesel were to become popular, it would have a tremendous impact on world demand, agricultural commodity prices, and the environment. The future market for biodiesel in China is potentially enormous.

Japan

Japan is totally dependent on imported, mainly Middle Eastern oil. Not surprisingly, in 1995 a three-year study was instituted to look into the feasibility of biodiesel production and use in Japan. A small pilot plant was constructed that relied on used frying oil collected from the Tokyo area as feedstock. In 1997 Japanese businessman Soichiro "Sol" Yoshida contracted with Pacific Biodiesel in Hawaii to design and build a biodiesel plant that would use waste cooking oil from his Kentucky Fried Chicken franchise in Nagano, Japan. The plant, which also processes used cooking oil from sixty other restaurants, produces 600,000 liters (132,000 gallons) of biodiesel per year and was designed so its capacity could be doubled.[20] The fuel produced is used in numerous cars, trucks, and industrial engines. Biodiesel was also used in buses during the 1998 Winter Olympics in Nagano.

Another series of biodiesel pilot projects have been conducted in Kyoto, where more than two hundred city-owned trucks have been fueled with B100 and eighty-one municipal buses have been operated successfully on B20. In 2001 the Shizuoka Trucking Association, composed of about 1,441 trucking companies, began buying farmers' entire canola (rapeseed) harvests in Shizuoka prefecture's Daitō and Iwata. The canola, which had been grown on idle plots of land, was then processed into biodiesel and mixed with petrodiesel for use in trucks. The association has been collecting performance data and has received numerous inquiries from other trucking companies that are also interested in using biodiesel.[21]

In 2002 a company in Japan installed what is believed to be the world's first co-generation turbine fueled with biodiesel. The system provides power and heat to the Matto-Ishikawa Central Hospital in Matto City, located about 400 miles northwest of Tokyo. The micro-turbine, originally made by Capstone Turbine Corporation in the United States, was modified to burn biodiesel.[22]

In December 2002 the Japanese government unveiled a comprehensive plan to encourage the greater use of biomass fuel that included the production of biodiesel from rapeseed and used frying oil. The plan also called for the creation of a biomass energy promotion council that would, among other things, lay out specific proposals for the construction of biomass energy production facilities.[23] More and more companies, municipalities, groups, and individuals across Japan are collecting used cooking oil and converting it into biodiesel for their own vehicle use or for sale to others. It is estimated that Japanese restaurants and households generate about 400,000 tons of used cooking oil every year.

As biodiesel use has increased, additional sources of feedstock have been investigated, and some researchers have turned to sunflower oil. Tsukuba University, in cooperation with the town of Hikawa, has been conducting sunflower-oil field tests, and in August 2003 a tractor fueled with sunflower biodiesel was demonstrated at the Hikawa Sunflower Festival.[24] Others have turned to canola (rapeseed) and are growing and processing their own canola oil in "nanohana projects" (*nanohana* means "canola blossoms" in Japanese), what has to be the ultimate grassroots recycling program. This is how it works. Local, small-scale community cultivation of rapeseed in crop rotation with rice is encouraged. Cooking oil is pressed from the harvested rapeseed, and after it is used for food preparation, it is collected by local housewives and converted into biodiesel. The pressed seedcake is used for fertilizer or animal feed. The biodiesel is consumed in the community. All of this happens at the local level in a closed loop. A national organization, the Nanohana Network, promotes the initiative and even holds an annual "Nanohana Summit."[25]

Although Japan has limited domestic sources of biodiesel feedstock, they will undoubtedly be utilized in a very efficient manner in the years to come. Many Japanese have embraced biodiesel enthusiastically, and future prospects for additional growth of the biodiesel industry are good.

Philippines

Coconut oil was reportedly used as an engine fuel in the Philippines during World War II. Research into the use of biofuels was later conducted in the 1970s as a result of the OPEC oil crisis. A program for biodiesel and one for bioethanol emerged from the research, but both programs were abandoned in the mid-1980s due in part to low oil prices.

The Philippines claims to be the world leader in coconut production. Every year the country exports about $760 million worth of coconut products and consequently has easy access to an enormous supply of biodiesel feedstock. Despite this enormous potential and the early research programs, it wasn't until November 2002 that the Philippines Department of Environment and Natural Resources (DENR) finally launched Esterol, a biodiesel made from coconut oil. In a practice-what-you-preach move, the department also initiated a policy of using biodiesel to fuel its own fleet of vehicles. Senbel Fine Chemicals Co. Inc. supplied the biodiesel, and the oil company Flying-V agreed to sell the fuel in all of its filling stations throughout the country. Since 2002 biodiesel blends have also been used extensively in a fleet of buses in Manila.[26] The Department of Agriculture assembled a team of experts from the Philippine Coconut Authority, the Technological University of the Philippines, and the Department of Energy to advance the Coconut Methyl Ester Program. In February 2004 a presidential order was signed directing all government agencies, including government-owned and government-controlled corporations, to switch to coconut methyl ester in their diesel-powered motor vehicles. The biodiesel industry in the Philippines appears to be well positioned for considerable future growth.

Malaysia

As mentioned in chapter 3, oil palm is the king of biodiesel feedstocks in terms of oil yield. And Malaysia claims to be the world's largest producer and exporter of palm oil products. It should come as no surprise, then, that the palm oil industry in Malaysia has experimented with biodiesel for some time. The Palm-Oil Research Institute Malaysia (PORIM) has been the prime mover in this initiative, and their 10,000-ton-capacity biodiesel pilot plant has been the main fuel source for research and field tests. Between 1987 and 1990 PORIM, together with Mercedes-Benz, conducted palm ester tests with city buses in Kuala Lumpur, with success. Additional successful tests were also run, with hundreds of trucks participating. Although palm oil biodiesel does not perform well in cold climates because of its high cloud point, there were no problems under the relatively hot local temperatures.

Despite all the research and testing, the biodiesel industry in Malaysia has not yet expanded to an industrial scale.[27] Malaysian government officials have recently urged Malaysian companies to become involved in joint ventures with foreign investors, particularly in Japan, where biodiesel feedstocks are limited, but the potential consumer market is huge.

Indonesia

In an effort to reduce air pollution in Jakarta, the city's administration announced plans in January 2004 to develop biodiesel fuel as an alternative to petrodiesel. The city will collaborate with the Riau provincial administration, which has a small biodiesel plant that will provide fuel for the test. The Jakarta Environmental Management Agency (BPLHD) and PT Energy Alternatif Indonesia, a biodiesel supplier, had previously conducted a joint experiment on ten public buses in the Indonesian capital using biodiesel blends ranging between B5 and B10. The initial tests were successful, and the subsequent larger, more sustained road tests were the result.

There are reportedly more than fifty types of plants that could provide biodiesel feedstock in Indonesia, and there is a good deal of potential for the industry to expand.[28]

Papua New Guinea

In November 2003 it was reported that Papua New Guinea intended to develop coconut oil as a biodiesel feedstock. The development of coconut-oil biodiesel is viewed as a high priority by the Papua New Guinea government and its various departments. The University of Technology in Lae has already made a good start with the initiative. The Biofuel Research Group at the university carried out a successful test of its coconut oil methyl esters the previous November. The initial results were promising and demonstrated that the fuel performs well.[29] Although it is just barely getting started, the biodiesel industry has good potential in Papua New Guinea.

Australia

Australia's widely dispersed population, in part, accounts for the nation's dependency on fossil fuels. Renewable energy, mainly from biomass recovered from the sugarcane industry, accounts for about 6 percent of Australia's total energy consumption. Nevertheless, the country has considerable potential to grow industrial oilseed crops on unused agricultural lands. Following the Gulf War in 1991, there was a brief flurry of biodiesel activity in Australia, but it soon waned. Despite the lack of interest at the national level, a number of dedicated biodiesel activists persevered, and finally they decided they needed an organized voice to represent them. As a result, the Biodiesel Association of Australia was founded in late 2000 to ensure a healthy biodiesel industry that represented the interests of all producers, from small backyard operations to industrial-scale plants. The organization now has more than five hundred members, and counting.

In January 2002 the Australian Biodiesel Consultancy, in collaboration with Collex, one of Australia's leading environmental services companies, began operation of a pilot plant in Wyong, New South Wales. The plant used recycled cooking oil and tallow as feedstock and supplied fuel for trucks, buses, cars, earth-moving equipment, and boats. On March 12, 2003, Australia's first commercial-scale biodiesel production facility, operated by Biodiesel Industries Australia, was opened in Rutherford, New South Wales. The multimillion-dollar plant, which has an annual capacity of 40 million liters (10.5 million gallons, or 37,000 metric tons), can use virgin vegetable oil, animal fats, or used frying oils for feedstock. The unique modular design of the production system allowed the plant to be set up in less than three months.[30]

Australian Renewable Fuels Pty. Ltd., a subsidiary of Amadeus Energy Ltd. of Perth, is planning a commercial-scale biodiesel facility in Adelaide, South Australia, followed closely by a second plant in Picton, Western Australia. The feedstocks for the plants will be animal fats and the continuous-process production technology will be provided by ENERGEA Umwelttechnologie GmbH of Klosterneuburg, Austria. The proposed facility in Adelaide will produce 45 million liters (11.9 million gallons, or 42,000 tons) of biodiesel annually, making it the largest biodiesel facility in Australia to date. Construction was scheduled to begin in 2004.[31]

In September 2003 an Australian biodiesel standard was finally implemented. Additional standard provisions will be gradually introduced over a number of years to allow existing producers time to get up to speed with testing and accreditation. What's more, the government hopes that biodiesel and bioethanol produced from renewable resources will contribute at least 350 million liters (93 million gallons) toward the nation's fuel supply by 2010. Future prospects for the Australian biodiesel industry are excellent.

New Zealand

Various energy crops were investigated in New Zealand in the late 1970s and early 1980s but were found to be noncompetitive with fossil fuels at the time. However, there are *a lot* of sheep and cattle in New Zealand, and the meat industry produces about 150,000 tons of tallow every year. A number of years ago, the Liquid Fuels Trust Board conducted research into the production of biodiesel from tallow. It was determined that about 10 percent of the nation's diesel fuel needs could be met from this source.[32] Other studies concluded that rapeseed would probably be the other best potential biodiesel feedstock for the country.

Recently, the New Zealand Energy Efficiency and Conservation Authority (EECA) has developed plans to use 120,000 tons of animal fat as biodiesel feedstock. The fat would be transformed into biodiesel in a new production facility that would be the first commercial-scale operation of its kind in the nation. A Massey University team is collaborating on the project. A number of companies are seriously looking into the possibility of using biodiesel made from tallow in their operations. Meridian Energy, the largest state-owned electricity generator in New Zealand, is one of those companies. Meridian would like to use biodiesel to power the excavating equipment used for its huge Project Aqua on the lower Waitaki River on South Island. The $1.3 billion project involves the construction of six power stations along a 60-kilometer canal on the south bank of the river.[33] If the biodiesel plan is implemented, there will be a large number of sheep-fat-powered bulldozers, scrapers, and dump trucks operating in New Zealand in the very near future.

Brazil

Brazil, the pioneer in promoting the use of biodiesel in South America, has to import more than 80 percent of the petroleum it consumes, leaving the nation at risk in the event of disruptions of foreign oil supplies. During the oil crisis of the 1970s, the rising price of oil contributed to Brazil's foreign debt crisis and led to the stagnation of the economy in the 1980s. In response to the oil crisis, there were reportedly some early engine tests with biodiesel conducted in the 1980s at Motores MWM Brazil in Sao Paulo, as well as a few road tests, but the research was subsequently discontinued. Nevertheless, Brazil is a huge consumer of biofuels, but the big seller is ethanol rather than biodiesel. Roughly 40 percent of Brazilian automobiles operate on 100 percent ethanol. The remaining cars run on a 22 percent ethanol and 78 percent gasoline blend, according to the U.S.-based Renewable Fuels Association. Annually, Brazil consumes nearly 4 billion gallons of ethanol fuel, which is produced there from sugarcane. Despite the focus on ethanol, there is a substantial opportunity for biodiesel to become a significant part of Brazil's renewable energy mix, since trucks and long-distance buses consume 90 percent of the petrodiesel sold in the country, amounting to 42 percent of all petroleum fuels, according to government statistics.

The Brazilian government, which had previously focused much of its attention on ethanol, now considers biodiesel to be a top priority. Beginning in 2002, the government initiated a new biodiesel fuel research program and also developed plans to offer financing and other fiscal incentives to help jump-start the industry. One proposal under discussion is to mandate a B5 blend for all diesel fuel sold in the nation, which might come into effect as soon as 2005. If the mandate were implemented, it would require 300,000 tons of vegetable (soy) oil to produce the biodiesel. That should not be a major problem, since the country produces about 4 million tons of soy oil annually and is second only to the United States in soybean production.[34] What's more, Brazil's huge ethanol industry can easily supply the alcohol needed for the transesterification process to make the biodiesel. A new biodiesel formula developed recently by researchers at the University of Sao Paulo (USP) in Ribeirão Prêto uses ethanol instead of methanol as the alcohol component, making the resulting biodiesel completely renewable.[35] Brazil's first commercial-scale biodiesel plant, Ecomat PR & Partners, located in Cuiabá, Mato Grosso, has an installed capacity of about 14,000 tons a year.

There have been a number of urban biodiesel road tests in Brazil. In 1998, and again beginning in 2001, the world-famous bus fleet in Curitiba, in the southern Brazilian state of Paraná, was involved in tests of B20 and other blends. Not only have the tests worked well in the buses, they also resulted in substantial reductions in the city's overall air pollution. In the first half of 2002, Rio de Janeiro's military police and electric utility vehicles were fueled with a mixture of biodiesel, petrodiesel, and alcohol. Additional trials with six city garbage trucks were conducted in 2003. The biodiesel feedstock for the tests was used frying oil from local McDonald's fast-food restaurants, prompting some locals to refer to it as "McDiesel." Based on this and other tests, biodiesel is now being considered for a wide range of uses.

Eletrotras (the main Brazilian electric utility) has announced a program that will promote the use of biodiesel as an alternative to petrodiesel at diesel-powered electrical-generating plants in the Amazon. The project calls for the local utility to gradually replace the petrodiesel used in ninety-one isolated systems with biodiesel made from mamona (castor oil plant) and palm oil. The government of the state of Rio Grand do Sul is particularly interested in using mamona as a biodiesel feedstock. The Farming Cooperative Mourãoense Ltd., one of the largest farming enterprises in Brazil, with seventeen thousand rural producers, plans to install a biodiesel production facility close to its headquarters in Paraná. The pilot unit will produce 2,500 liters (1,650 gallons) of biodiesel a day from soy oil. The fuel will initially be used in the cooperative's vehicles.[36]

Finally, as mentioned previously, some Brazilian trains in the southern part of the nation are on track to expand biodiesel use even further. In December 2003 the transportation company América Latina Logística (ALL), with 15,000 kilometers (9,321 miles) of railroad lines in southern Brazil and Argentina, decided to replace a quarter of the petrodiesel it consumes (150,000 tons per year) with biodiesel. Preliminary tests were conducted on two trains in early 2004, and there are plans to expand biodiesel use to the entire system. If that occurs, ALL will use 35 million liters (32,000 tons) of biodiesel a year in its 580 trains, "enough to make a fuel production plant feasible," says Antonio Tomasi Filho, coordinator of the company's supply division and one of the heads of the project.[37]

The future prospects for the biodiesel industry in Brazil are excellent, especially if the nationwide B5 mandate is implemented.

Paraguay

Paraguay imports all of its petroleum and petroleum products. Disruptions to international oil supplies as well as the upward trend in oil prices in recent years have given the Paraguayan government a strong incentive to investigate the use of biodiesel. In October 2000 the Vice-Minister of

Mines and Energy announced a biodiesel initiative and the government officially launched studies to verify the technical and economic feasibility of the fuel. As part of the initiative, Hardy S.A., an oleochemical company based in Asunción, began to produce small quantities of biodiesel. The government was particularly impressed with the fact that biodiesel technology is relatively simple and can be scaled to meet the needs of local communities. Subsequently, a number of government agencies began to travel throughout the country to explain the initiative and set up pilot projects and trials. Soybeans are the principal biodiesel feedstock in Paraguay, and a national B5 biodiesel blend would require the production of about 50,000 tons of soybean oil every year. Ethanol produced from sugarcane is the principal source of alcohol for the transesterification process.

In October 2001 the first collaboration with the municipal government of the capital city of Asunción began, and a number of road tests were conducted. In September 2002 B30 was introduced at local filling stations at a price lower than that of petrodiesel. In the same month a bus company, Linea 25, began using a biodiesel blend in all of its buses, and production at the Hardy S.A. plant reached 5,000 liters (1,320 gallons) per month.[38]

The U.S.-led war in Iraq in 2003 and the following uncertainties in the international oil market have bolstered the Paraguayan government's resolve to place even more emphasis on biodiesel to strengthen its energy security. The creation of a National Biodiesel Network to coordinate the activities of the various players has been proposed, and expansion of biodiesel production facilities as well as the number of biodiesel users is expected.

Argentina

Although Argentina has put most of its recent alternative-energy emphasis on compressed natural gas, there has been some biodiesel activity in the country. As of 2001, there were at least fourteen small-scale biodiesel production projects in operation in various locations. Most of these biodiesel production facilities are reportedly still operating, and a national biodiesel fuel standard has been adopted.

Other South American Countries

Colombia reportedly has planned a biodiesel plant with a 60,000-gallon (210-ton) capacity for Bogotá that will supply biodiesel to the internationally famous sustainable-living community of Gaviotas. Palm oil will be the primary feedstock.[39]

There are a number of small biodiesel projects scattered throughout many of the other nations of South America, but to date they have not reached large-scale commercialization.

Nicaragua

A successful study on the use of jatropha oil as a biodiesel feedstock by the National Engineering University (UNI) in Managua, Nicaragua, supported by an Austrian development program, was completed in 1996. This was followed by the construction of the country's first biodiesel pilot plant as well as a quality-control laboratory. The facility, which opened in 1997, has an annual production capacity of 3,000 tons. In addition to the construction of this facility, 1,000 hectares (2,471 acres) of previously degraded and idle agricultural land was planted with *Jatropha curcas* and was expected to yield about 426,000 gallons of oil after the bushes matured in about five years (the plantation has reportedly been expanded to 1,500 hectares). The plantation was expected to supply enough oil to meet about 3 percent of the country's total annual petrodiesel consumption. The biodiesel initiative is part of the larger Biomass Project, a collaboration between UNI, the Nicaraguan Ministry of Energy (INE), and the Austrian firm of Sucher & Holzer.[40] Also, a biodiesel filling station has reportedly opened in Managua.

Other Central American Countries

Costa Rica, which spends about $500 million on petroleum fuels every year, appears to be on the verge of embarking on a national biodiesel initiative. At the end of March 2004 the Costa Rican Center for Cleaner Production announced that it would present a proposal for the wide-scale production and use of biodiesel created from natural oil sources such as the African palm.[41]

A number of isolated, small-scale biodiesel operations are reportedly scattered across most of the other nations of Central America.

Canada

Canada's Alternative Fuels Act was implemented on April 1, 1997. The main purpose of the act was to increase the use of alternative transport fuels (ATF) in Canada by gradually increasing the percentage of vehicles in government fleets capable of operating on alternative fuels to 75 percent by 2004. Although those targets have been met, Canada has been slow to adopt the use of biodiesel. This may, in part, be due to concerns about the fuel's performance in the country's notoriously cold winters. Nevertheless, Agriculture and Agri-food Canada recently reported that if the world's thirty major economies replaced just 8 percent of the fossil fuel they consumed with biofuels, commodity prices would rise enough to solve the farm income crisis that has bedeviled the agricultural sector in recent years. Canadian farmers have been taking note, and in the past few years biodiesel activity in Canada has been heating up.

In April 2001 the BIOX Corporation started biodiesel production in a demonstration plant in Hamilton, Ontario, capable of producing 1 million liters (927 tons). The multifeedstock process for the plant had been developed earlier at the University of Toronto. Biox plans to open a commercial-scale biodiesel plant with a capacity of 56,000 tons per year based on this technology in 2005. Another company, Rothsay, has been producing biodiesel in their Montreal plant since September 2001 using restaurant oil and animal fat as feedstocks. The 4-million-liter-capacity, batch-process plant was designed by Rothsay, and the company is upgrading the facility to produce 35 million liters. The $14.5 million plant is expected to be completed in the spring of 2005. Rothsay is a fully owned subsidiary of Maple Leaf Foods Inc. of Toronto.

In March 2002 the city of Montreal began testing biodiesel in some of its municipal buses, and by midsummer 155 Montreal Transit Corp. buses on nineteen routes serving the downtown area were running on B20. The biodiesel was made from used vegetable oil and animal fats. Although the program was a technical success, it was discontinued the following summer when the Quebec provincial government refused to extend a tax break of 16.4 cents per liter (62 cents per gallon). In June 2002, however, the province of Ontario initiated a tax exemption of 14.3 cents per liter for biodiesel, and the city of Brampton, Ontario, became the first municipality in Canada to commit to the regular, ongoing use of biodiesel. That summer, the city began using biodiesel in its fleet vehicles, and in October all of its 137 buses began running on B20.[42] Other bus systems in Canada that are using or testing biodiesel include those of Guelph and Sudbury, Ontario, and Saskatoon, Saskatchewan. In 2002 electric utility company Toronto Hydro switched its entire four-hundred-vehicle fleet to B20. To date, thousands of fleet vehicles have driven millions of kilometers using biodiesel in Canada.

As an indication that the Canadian biodiesel industry is beginning to develop momentum, a new organization composed mainly of oilseed growers and processors, Biodiesel Canada, met for the first time in May 2003 in Winnipeg, Manitoba. The group has already begun to work with the government on a variety of issues related to oilseed production and fuel taxes. And on March 2, 2004, Canada's first public biodiesel fuel pump opened in Unionville, near Toronto. The B20 blend for the pump was supplied by Ottawa-based Topia Energy Inc., which was the first company to open a Canadian biodiesel terminal (also in the Greater Toronto area).[43] In May 2004 a second public pump was opened by Topia, this time on Queen Street in Toronto. In addition, Topia is actively pursuing plans to build a large-scale commercial biodiesel production plant in an eco-industrial park located in Sudbury. The first phase of the plan calls for a modular facility that will produce 3 million liters (2,770 tons) annually. Production will be increased in a series of steps as more modular units are added later. Not surprisingly, the agricultural lands adjacent to Sudbury represent one of the best canola (rapeseed) growing regions in Ontario. Sudbury has a growing reputation for being interested in alternative energy enterprises. The city's highly publicized wind power project was instrumental in establishing the contacts that brought the biodiesel proposal to the community.[44]

Overall, the Canadian biodiesel industry has substantial potential for future growth.

PART THREE

Biodiesel *in the* United States

8

A Brief History

With only 4.5 percent of the world's population, the United States consumes about 25 percent of global energy and produces roughly 25 percent of the planet's CO_2 emissions. Because of this dubious distinction, the opportunities for positive change in U.S. energy practices are enormous—and the need couldn't be more urgent. The United States presently imports about 60 percent of its oil, and that figure is going to rise in the years ahead. Americans collectively spend about $200,000 per minute to pay for that imported oil. This enormous outflow of money not only contributes substantially to the U.S. balance of payments deficit but also unquestionably has helped fund at least some anti-American activities around the globe.

Unfortunately, many U.S. politicians, especially those who spend most of their time in Washington, D.C., still don't get it. They have repeatedly resisted simple but important steps such as improving federal fuel economy standards, failing to understand that the consequences of inaction are becoming increasingly costly and far outweigh the short-term political advantages of maintaining the status quo. If there was ever a place and time for courageous and visionary political leadership regarding renewable energy, the place is the United States, and the time is now. Because it plays such a key role in the energy dilemma currently facing the planet, anything the United States does to try to clean up its act will have a significant impact on the rest of the world. This is why the next four chapters are devoted entirely to the biodiesel industry in the United States. First, a little background.

Early Biofuels

Although biodiesel is a relatively recent development in the United States, the basic chemistry involved has its roots in the nineteenth century. As early as the mid-1800s, transesterification was used as a strategy for making soap. Early feedstocks were corn oil, peanut oil, hemp oil, and tallow. The alkyl esters (what are now called biodiesel) resulting from the process were originally considered just by-products. Ethanol also had a significant place in early U.S. biofuels history. Prior to the Civil War, an ethyl alcohol (ethanol) mixed with turpentine, known as camphene, was widely used as a lamp oil. A tax on alcohol enacted during the Civil War dramatically reduced the use of industrial alcohol, and it was not until 1906, when the tax was finally repealed by Congress, that ethanol began to make a comeback as a fuel, especially for internal combustion engines.[1]

A number of Europeans are credited with inventing the internal combustion engine, especially Nikolaus August Otto, who produced an early version around 1866. However, Samuel Morey of Orford, New Hampshire, built the first prototype internal combustion engine in the United States in 1826. He used a biofuel, alcohol, as the main fuel in his experiments. Unfortunately Morey, who was many decades ahead of his time, was unable to attract financial backing for his invention, and he has been largely ignored by historians. As noted previously, Rudolf Diesel was also a firm believer in the potential of biofuels for powering his engine. But Diesel was not alone in his enthusiasm.

In 1896 Henry Ford built his first automobile, a quadricycle, to run on ethanol. In 1908 Ford's famous Model T was designed to run on ethanol, gasoline, or a combination of the two. Ford was so convinced that the success of his automobile was linked to the acceptance of "the fuel of the future" (as he described ethanol) that he built an ethanol production plant in the Midwest. He then entered into a partnership with Standard Oil Company to distribute and sell the corn-based fuel at its service stations. Most of the ethanol was blended with gasoline. Ford's biofuel turned out to be fairly popular, especially with farmers, and in the 1920s ethanol represented about 25 percent of Standard Oil's fuel sales in that part of the country. In retrospect Ford's alliance with Standard Oil may not have been such a good idea. As Standard Oil tightened its grip on the industry, it focused its attention on exploiting its petroleum markets—and eliminating any competition. Nevertheless, Ford continued to promote ethanol through the 1930s. But finally, in 1940, he was forced to close the ethanol plant due to stiff competition from lower-priced petroleum-based fuels.[2]

During World War II, because of the disruptions to normal oil supplies, virtually all the participating nations made use of biofuels to power some of their war machinery. At the same time, transesterification used in the soapmaking process became the subject of great interest because of the by-product glycerin, which is a key ingredient in the manufacture of explosives. But after the war, the return of steady oil supplies and low gasoline prices brought an end to biofuels production in the United States. Some observers maintain that the demise of the biofuels industry in the 1930s and '40s was due to the deliberate actions of a small group of individuals such as William Randolph Hearst, Andrew Mellon, the Rockefellers, and a number of "oil barons." Others insist that the biofuels industry was simply the victim of larger market forces. In either case, for all practical purposes the industry ceased to exist after World War II. It was not until the oil shocks of the 1970s that biofuels began to experience a renaissance.

The OPEC Oil Crisis

The oil crisis of the 1970s was a rude awakening for most Americans, dramatically underscoring the nation's dependency on imported oil. On October 17, 1973, when the OPEC nations shut off the petroleum spigot to the West, oil imports from Arab countries to the United States quickly dropped from 1.2 million barrels a day to a mere trickle. As the price of oil increased dramatically and long lines at gasoline stations grew even longer, people across the country began to look for alternative sources of energy. In 1973 President Richard M. Nixon created a cabinet-level

Department of Energy headed by William E. Simon, who became known as the nation's "energy czar." Many of the early federal renewable energy programs at agencies such as the National Renewable Energy Laboratory (formerly the Solar Energy Research Institute) were also initiated during the late 1970s.

The 1979 revolution in Iran that resulted in the ouster of the U.S.-backed Shah precipitated yet another global energy crisis. Oil prices doubled, sending the industrial world into a recession. When the crisis finally subsided, oil prices fell, and the Reagan Administration ended the tax incentives and other support for the renewable energy industry. The initial steps that had been taken toward a comprehensive national renewable energy initiative were largely abandoned.

Early Experiments

Despite the loss of federal support, some of the enthusiasm for renewable energy lingered on and even began to grow. As mentioned in chapter 2, some of the earliest experiments into the production and use of biodiesel in the United States took place at the University of Idaho, beginning in 1979 with the work of Dr. Charles Peterson. For the first few years Peterson's experiments were relatively low-key and focused mainly on farm tractors. It wasn't until the late 1980s and early 1990s that the scope of his research began to expand exponentially. But Peterson was not the only one involved in biodiesel research.

Colorado

In the summer of 1989 Dr. Thomas Reed, who was on the faculty at the Colorado School of Mines, first learned about the conversion of animal fats and vegetable oils into biodiesel. Reed, who had a long-time interest in biofuels, began to wonder whether used cooking oil would make a viable feedstock. In November, after researching the available literature (especially Dr. Peterson's), he decided to conduct his first transesterification experiment. "I got a gallon of used vegetable oil from the grease dumpster at a local McDonald's," he recalls. "Then, I brought it into my lab, made some minor adjustments in the methanol and lye recipe, and stirred it up well, and when I got back from lunch, there it sat—a gallon of beautiful fuel from that awful grease. I couldn't believe how easy it was."[3]

Reed was so enthusiastic that he continued his transesterification experiments with a variety of used oils, even making biodiesel from bacon grease on Christmas in his daughter-in-law's kitchen. In the spring of 1990, Reed approached the Denver Regional Transportation District (RTD) to see if they were willing to try some biodiesel in one of their buses. They agreed, but first Reed had to make use of a larger laboratory at the School of Mines in order to produce the 100 gallons of biodiesel needed for the tests. After the fuel had been made, tests were conducted on an RTD bus, and they showed substantial reductions in emissions at various percentages of biodiesel. The remainder of the fuel was used to operate a bus in Denver for about a week. This may have been the first public test of biodiesel in a commercial vehicle in the United States.

Reed decided that "transesterified waste vegetable oil" was not a very catchy name, and, considering the source, he decided to call it McDiesel. "I applied for a copyright," he says. "I even approached McDonald's to see if they were interested. They were, but said they would sue me if I used that name. Later people came to call these fuels 'biodiesel,' and I now live with that. However, I would love to have had McDonald's sue me—what publicity!" With two other partners, Reed formed a company to make and promote his fuel. They published a number of papers, held numerous discussions with many people, and finally entered into an agreement with a chemical engineering company. But when the agreement with that company descended into protracted legal wrangling, Reed lost his enthusiasm for biodiesel and moved on to other renewable fuel interests (more on Dr. Reed a little later).[4]

Missouri

At the University of Missouri at Columbia, Leon Schumacher, an associate professor in the Agricultural Engineering Department, also became involved in the second wave of biodiesel research that began in the early

1990s. Schumacher was approached by members of the Missouri Soybean Merchandising Council in 1991 about conducting research on new uses for soybean oil. He was intrigued. Schumacher studied reports of earlier research, but he wanted to focus on a project that had not already been conducted in the United States. "An acquaintance of mine pointed out that most of the early testing had been done on farm tractors, but that no one had done much with the use of biodiesel as a fuel in regular diesel-powered vehicles like cars and trucks," Schumacher recalls. That was what he had been looking for. Schumacher applied for the Soybean Council project.[5]

With financial assistance from the Agricultural Experiment Station at the university, a new 1991 Dodge diesel pickup track was purchased. "We put signs on the side of the truck that said 'Fueled by 100 percent SoyDiesel,' and it wasn't long before it began to catch the eye of a lot of people," Schumacher says. "I distinctly remember one day there was a guy riding a motorcycle who was passing me, and he looked at the sign on the side of the truck and took his hand off the throttle, made a fist, and raised his thumb as a sign of approval. There was a lot of that kind of public acceptance and support." The biodiesel used in the test came from Midwest Biofuels, a subsidiary of Interchem Environmental Inc. headquartered in Overland Park, Kansas. The company's small-scale production facility was actually located just across the state line in Kansas City, Missouri. "I think the first 100 gallons they produced for us were made in garbage cans in the backyard," Schumacher says, laughing.

Before long the truck was on the road almost constantly, and it was used as part of an education and promotion initiative for farmers, legislators, and the general public. The truck ran on B100 for 90,000 miles before the test was ended in 1996. During that time, fuel consumption was carefully monitored and dynamometer and other tests were conducted. "One thing we noticed was that there was virtually no soot coming out of the exhaust pipe, and that really caught our attention," Schumacher says. In 1992 a new Dodge pickup truck was purchased for a second test and it ran for 100,000 miles on B100. Then the engine was removed and sent to the manufacturer in Indiana, where it was torn down and carefully inspected. The engine was found to be in excellent condition.

Over the years Schumacher has been involved in an eclectic series of biodiesel experiments with garbage trucks in Columbia, buses in Saint Louis, and even a diesel locomotive in Saint Joseph, Missouri. For the locomotive test Schumacher collaborated with Jon Van Gerpen, an associate professor from the Mechanical Engineering Department of Iowa State University in Ames, Iowa, who has headed UIA's biodiesel program from its inception. The locomotive, a veteran 1945 GM diesel-electric switcher, was being used to move grain cars for the Bartlett Grain Company. The test was designed primarily to measure the amount of exhaust smoke produced with a B20 biodiesel blend. A light gray smoke was observed during engine acceleration, but no differences were noted by the locomotive operators concerning fuel economy, engine oil consumption, engine oil dilution, or fuel compatibility. "We had to install a special battery power unit for our testing equipment because there wasn't anything we could plug into on the locomotive," Schumacher recalls.

Schumacher is quick to point out that there were other individuals at the University of Missouri and many other institutions as well who were also involved in various types of biodiesel research. "We worked with West Virginia University on the Saint Louis bus project, and we often collaborated with the National Renewable Energy Laboratory in Golden, Colorado," he says. "We shared quite a lot of our data with them, and it became part of a more comprehensive study because they were looking at a number of other fuels as well." Additional institutions involved in substantial biodiesel research have been Iowa State University at Ames, Kansas State University, the University of Illinois at Champaign-Urbana, North Dakota State University, the University of Tennessee, and the University of Georgia, as well as numerous soybean growers' associations from Midwestern states.

Schumacher's most recent biodiesel research project has been with the South Dakota Department of Transportation, which is investigating a B5 blend for use in its vehicles. "I feel very good about it," Schumacher responds, when asked about his thirteen-plus years of biodiesel research work. "If I had it to do all over again, I might do a few things a little differently, but I guess that's normal. Each project has had its unique challenges, but they've all been interesting."[6]

Early Production

Although there was unquestionably some biodiesel research taking place here and there across the country in the 1980s and early 1990s, most of the national biofuels emphasis at the time was focused on ethanol. But around 1990, that began to change. Some of the earliest U.S. production of biodiesel in commercial quantities took place in the early 1990s at the nondedicated plant of Procter & Gamble in Kansas City, Missouri.

Midwest Biofuels

In 1991, Midwest Biofuels, a subsidiary of Interchem Environmental Inc., set up a small-batch-process pilot plant in Kansas City, Missouri, to supply limited quantities of biodiesel for Leon Schumacher's SoyDiesel-powered Dodge pickup truck from the University of Missouri and similar demonstration projects. "I remember making the stuff by hand," recalls Bill Ayres, one of the company's founders. "Then Leon drove up in this brand new pickup truck with its twin fuel tanks. We filled up one of the tanks with B100, he flipped a switch, and from that point on the truck ran for many years on SoyDiesel."[7]

The following year the pilot plant also supplied biodiesel for the first phase of a trial at Lambert Airport in Saint Louis involving about ten vehicles. "I didn't have the facilities to produce a lot of biodiesel at the time, and I was a little worried about the gel temperature in the winter, so, more or less out of the blue, I suggested a B20 blend," Ayres says. B20 subsequently became a standard for many other trials and is the most popular blend in use today across the United States. As demand for biodiesel for additional trials and tests grew, Midwest BioFuels set up a new, larger plant in Kansas City, Kansas, in 1993 with a production capacity of 100 gallons per batch (later expanded to 500 gallons). This was the first dedicated commercial biodiesel facility in the United States. But after about six months, the demand started to outstrip Ayres's ability to meet it, and he began to purchase biodiesel from the nearby Procter & Gamble plant.[8]

Ag Processing Inc.

In October 1994 Ayres and a business partner, Doug Pickering, left Interchem to become consultants for Ag Processing Inc. (AGP), the country's largest soybean-processing cooperative. AGP has about two hundred thousand members in sixteen states and Canada. In the following year Ayres and Pickering formed a new joint venture with Ag Processing called Ag Environmental Products (AEP). In 1996 AGP opened a new batch-process biodiesel plant with a capacity of 5 million gallons (17,500 tons) in Sergeant's Bluff, Iowa, adjacent to an existing seed-crushing facility. In the spring of the following year AEP provided biodiesel fueling stations at ten farm co-op locations in six Midwestern states. Other stations were subsequently added. Over the years the $6 million soy methyl ester facility at Sergeant's Bluff has produced a wide range of products, including biodiesel, solvents, and agricultural chemical enhancers under the SoyGold brand name, which Ayres and Pickering promoted and marketed for AEP. Biodiesel produced by AGP has been used in a wide range of vehicles by customers across the country.

Twin Rivers Technology

In 1994 Twin Rivers Technology began business in a former Procter & Gamble oleochemical facility in Quincy, Massachusetts. The plant had a nominal production capacity of 30 million gallons, but the company never actually made any biodiesel at the Quincy facility, purchasing it instead from other plants. In 1996 Twin Rivers was the first company in the nation to receive U.S. Environmental Protection Agency certification of its Envirodiesel brand of biodiesel fuel for use in meeting Clean Air Act compliance under the agency's Urban Bus Retrofit Program. However, the company suffered from a long series of marketing and internal problems and closed its biodiesel division in 1998 (for more on Twin Rivers, see chapter 9).

NOPEC

In 1995 the NOPEC Corporation began initial production at its huge 18-million-gallon-capacity batch-process biodiesel plant in Lakeland,

Florida, using roughly 50 percent virgin soybean oil and 50 percent used cooking oil as feedstocks. Although NOPEC (the company name poked fun at OPEC) landed a large contract to supply 23,000 gallons of biodiesel to the Iowa Department of Transportation, the company initially focused mainly on the marine market. NOPEC's main customers were boat owners in the environmentally sensitive waters of the Florida Keys and Maryland's Chesapeake Bay. The National Oceanic and Atmospheric Administration's patrol boats in the Florida Keys National Marine Sanctuary were early users of biodiesel from NOPEC.[9] Other customers included the U.S. Postal Service and numerous school districts across the state. In 1997 and 1998 the company ran into serious legal problems related to the sale of unregistered stock, and in December 2000 the NOPEC production facility was purchased by OceanAir Environmental, a California-based company (for more on NOPEC, see chapter 9).

West Central Cooperative

In 1996, in response to increasing interest and demand, there was a sudden flurry of biodiesel plant construction across the country. West Central Cooperative, in the tiny community of Ralston (population 119), Iowa, built a new 2.5-million-gallon-capacity batch-process plant. The facility was a success and was subsequently expanded to increase its output. But there were some lingering inefficiencies with the expanded design, and in 2001 West Central decided to start with a completely new continuous-process facility. The new 12-million-gallon-capacity plant began production in 2002. "It's a state-of-the-art facility," says Gary Haer, who has been with West Central since 1998 (he was previously with Midwest Biofuels/Interchem). "We recover all of the materials that are used to convert the soybean oil into biodiesel, so there's no waste. Everything is recaptured and reused; it's a unique production facility."[10] West Central now has the largest exclusively soy continuous-process biodiesel plant in the United States.

Columbus Foods

The founding of West Central was followed in the fall of 1996 with the start-up of a biodiesel production facility in Chicago, Illinois, by Columbus Foods, a food-grade-oil producer and distributor for more than sixty years. Earlier that year the company's owner, Michael Gagliardo, had read about biodiesel in a magazine article and had been inspired by Chicago's demonstration of biodiesel in buses and police patrol boats. He was intrigued by the possibilities of producing a new product from used oils, and after traveling to Austria to learn more about it, he had a 200,000-gallon-per-year plant constructed at a cost of $500,000. The City of Chicago donated a building through its brownfield program, which was designed to clean up environmentally degraded urban lands. Early demonstration users for a Columbus Foods B20 blend were the Chicago Transit Authority and the American Sightseeing Bus Company.[11] But after experiencing some cold-weather difficulties with biodiesel made from used oil, the company decided to switch to using virgin soybean oil as their primary feedstock. With some additional equipment and newer processing techniques, Columbus Foods now can produce up to 6 million gallons a year.[12]

Pacific Biodiesel

The last biodiesel production facility to come online in 1996 was built by Pacific Biodiesel in Kahului on the island of Maui in Hawaii. Bob King had been a diesel engine mechanic for over twenty years and was the owner of King Diesel, which provided diesel engine service primarily for the marine industry. King, who had been contracted to maintain the diesel-powered electrical generators at the central Maui landfill, noticed that the landfill was being swamped with tons of used restaurant grease from trucks that serviced local hotels and restaurants. He grew increasingly concerned about the potential environmental and health problems caused by the grease, and in 1995 he decided to do something about it. He eventually contacted Daryl Reece, a researcher at the University of Idaho who had worked with Dr. Charles Peterson as both an undergraduate and a graduate student on ways of making biodiesel fuel from discarded cooking oil. King had found the answer to the grease problem—and a business partner as well.

King and Reece formed Pacific Biodiesel Inc., and without outside financing they proceeded to build the first biodiesel plant on the Pacific Rim and the first commercial facility in the United States to rely entirely on used cooking oil as its feedstock. "We couldn't get any normal financing because nobody had heard of biodiesel," King recalls. "And we didn't have a market for the product because no one knew what it was. It probably wasn't a very wise decision, but we didn't know any better at the time."[13] After the plant was completed at the Maui landfill site, the trucks that had previously pumped used cooking oils into the landfill found that it was more economical to deliver their loads to the biodiesel facility.[14]

Initially King's vision for the venture didn't extend beyond Hawaii. But that changed almost immediately. "When we held the grand opening for the first plant there were people who came from all over the place, and we got press coverage from all over the world," he says. "I quickly realized that this wasn't just something that was going to apply to Maui; there were some much larger implications to this." When the company set up its Web site, it began to receive inquiries from people in nations around the globe. The following year Pacific Biodiesel built a similar plant in Nagano, Japan, for a fast-food restaurateur (see page 132).

Shortly after the completion of the project in Japan, King and Reece turned their attention to an even larger problem at the Maui landfill—grease-trap waste (brown grease) from sewer grease traps in restaurants and other grease sources. Reece designed a custom processor for grease-trap waste, and Pacific Biodiesel was able to create its own boiler fuel while diverting 140 tons of grease-trap sludge from the landfill every month. Building on its successful operation on Maui, Pacific Biodiesel built a 1,500-gallon-per-day biodiesel plant in Honolulu that processed 25,000 gallons of grease-trap waste per day. The Honolulu facility now processes used cooking oil as well.

"It took a little bit of time to convince people that this fuel wasn't going to ruin their engine or cause problems, so it was a very slow, methodical, customer-by-customer movement of the product into the market for the first three years," King says. "We put a lot of time into education about what the product was. Then it started getting easier, but it has only been in the past year or so that biodiesel has become a common word and people finally seem to know what it is."[15]

The Soybean Factor

Unlike the European biodiesel industry, which was largely a creation of national-government-sponsored research and energy policies, much of which was focused on rapeseed, the biodiesel industry in the United States has had its longest and strongest support from Midwestern soybean farmers. The United States is the largest producer of soybeans in the world, with an average annual production of about 70 million tons. The nation is also the largest consumer and exporter of soybeans on the planet. Because the demand for soy meal has generally risen faster than the demand for soy oil for many years, the soybean industry has been faced with the dilemma of what to do with the surplus oil. Much of this surplus has been directed toward the biodiesel market, especially in recent years, which largely explains the enthusiasm that soybean growers and processors have had for the fuel. One bushel of soybeans is needed to produce about 1.5 gallons of biodiesel.

The American Soybean Association

It should come as no surprise that one of the organizations that took an early interest in biodiesel was the American Soybean Association (ASA) and its associated marketing, research, and communication arm, the United Soybean Board (USB). Originally founded in 1920, the ASA has worked for many years to promote the interests of soybean farmers. In 1991 the association implemented a national soybean checkoff of 0.5 percent on the per-bushel market price of soybeans when the crop is first sold to help fund market promotion, research, and educational programs. Two years later the ASA and the USB began to implement checkoff-supported programs, and the ASA became heavily involved in a wide range of legislative, research, and development activities for what they call soydiesel.

"We saw a wonderful opportunity to use a homegrown crop to meet our country's energy needs, and to do so in a way that helps farmers while improving air quality and making the country safer," says Neil Caskey, special assistant to the CEO of ASA. "Over the years, soybean farmers have invested about $35 million through their checkoff in building a biodiesel market. This is something that they believe in, and they are putting their money where their mouth is. This is one of the reasons why this initiative continues to grow and get a lot of support. Also, when people see that farmers are willing to use biodiesel on their own farms in their own trucks and equipment, they understand that they are truly committed."[16]

The National Biodiesel Board

In 1992 the USB decided to promote an American biodiesel project and founded the National SoyDiesel Development Board, which was succeeded in 1994 by the National Biodiesel Board (NBB). The NBB, headquartered in Jefferson City, Missouri, has developed into the national trade association representing the biodiesel industry and is the coordinating body for research and development in the United States. The NBB also works with a broad range of industry, government, and academic entities on a wide range of biodiesel-related activities. Members of the organization include state, national, and international feedstock-growing and processing organizations; biodiesel suppliers, marketers, and distributors; and technology providers—all big players in the industry. The NBB also supports quality assurance programs and acts as a clearinghouse for information.[17] The NBB Web site (www.nbb.org) offers a vast amount of constantly updated information about biodiesel and the industry.

Joe Jobe, the executive director of the NBB, has high praise for the farmers who have worked tirelessly to support the biodiesel industry. "The support of the soybean farmers has been absolutely critical," he declares. "If the soybean farmers had not decided to invest—through their state and national soybean checkoff programs—in biodiesel research and development programs, we wouldn't have a National Biodiesel Board. We wouldn't have a legally registered fuel designated as an alternative fuel, the only alternative fuel in the country to have complied with the health-effects testing requirements of the 1990 amendments to the Clean Air Act. We also wouldn't have one of the best-tested alternative fuels in the country, with more than 50 million successful road miles and countless off-road and marine hours in virtually every diesel engine type and application. Soybean farmers have driven the development of the U.S. biodiesel industry, both through their investments of their money and through their grassroots support of biodiesel."[18]

"The National Biodiesel Board is an exceptional organization that has supported our industry in so many ways," says Gary Haer of West Central Soy. "The resources and technical information that they provide to the industry and the general public are remarkable, and they have really assisted in the growth of this market. The soybean industry and the NBB have helped make biodiesel the success story that it is today."[19]

Other Organizations

There have been a number of other organizations that have supported the development of biodiesel over the years. The U.S. Department of Energy and a number of its affiliates, such as the Alternative Fuels Utilization Program and the National Renewable Energy Laboratory (NREL) in Golden, Colorado, have had biodiesel initiatives. NREL, in particular, has conducted research into many aspects of biodiesel, the most notable of which is the lab's extensive research into algae as a feedstock in Roswell, New Mexico. Less familiar organizations, such as the Fat and Protein Research Foundation (FPRF) and the National Renderers Association (NRA), have also benefited the biodiesel industry with their research efforts to develop new outlets for waste oils and fats from plants and animals.

Homebrew

In addition to the activities of soybean farmers and commercial producers, for many years there has been a parallel grassroots biodiesel movement in the United States propelled by a group of dedicated biodiesel enthusiasts. Like their counterparts in many other countries, these individuals grew tired of waiting for the government to establish a comprehensive national biodiesel initiative and went ahead and started to make and use biodiesel on their own. Quite a few small-scale producers scattered across the country have been making biodiesel from used cooking oil in kitchen blenders, old 55-gallon drums, and an assortment of other funky, low-tech pieces of equipment for many years.

The widely distributed homebrew feedstock sources, such as individual fast-food restaurants (especially those located in rural areas), do not lend themselves well to supplying large amounts of used oil to industrial-size biodiesel production facilities. But these scattered feedstock sources are a good match with equally scattered backyard "home brewers," who often produce just enough fuel for their own use or, perhaps, for a few close friends. While used cooking oil is a limited resource, it does offer biodiesel producers, especially smaller producers, an inexpensive (usually free) and readily available alternative to more expensive virgin vegetable oils. And producing fuel from local resources for local use makes a lot of sense from an energy-efficiency and environmental standpoint.

Fat of the Land

Many knowledgeable people in the homebrew sector credit Dr. Thomas Reed of Colorado (mentioned earlier) for developing and popularizing a simple, easy-to-follow recipe for making biodiesel in small quantities from used cooking oil. As his recipe spread by word of mouth, especially in the back-to-the-land and counterculture communities, more and more people began to experiment with making their own fuel from old grease. In 1993 independent San Francisco filmmaker Nicole Cousino and her team of documentary filmmaker friends, Sarah Lewison, Julie Konop, Florence Dore, and Gina Todus, discovered Reed and his recipe. "I first heard about running vehicles on vegetable oil a few years earlier from an inventor in upper New York State named Louis Wichinsky," Nicole Cousino recalls. "He was the one who put me in touch with Tom Reed at the Colorado School of Mines. I had never taken chemistry, so Tom and a friend of his,

Agua Dass, spent perhaps forty or fifty phone calls patiently explaining to me what we needed to do in order to make the fuel."[20]

In July 1994, armed with their newfound knowledge, the five women filmmakers took a pioneering grease-powered trip across the country from New York City to San Francisco in a GMC diesel van, stopping at an endless succession of greasy spoons and other restaurants, making biodiesel along the way, and interviewing a number of early biodiesel pioneers, including Dr. Reed (who graciously allowed them to make biodiesel in his laboratory). They also produced a humorous documentary about their adventure titled *Fat of the Land.* Although this classic 1995 cult film was unabashedly humorous, it nevertheless raised serious questions about the stranglehold that petroleum has on the U.S. economy, American energy usage, and sustainability. "Even today, the whole idea is still amazing; there is a certain radical element to being able to run your car on waste vegetable oil," Cousino says. "We also wanted to make the point that this was accessible science; that if we could do it, anybody could do it. Tom Reed often referred to this as 'bucket science' or 'kitchen science,' and it really is."[21] *Fat of the Land* is a must-see for biodiesel fans who want to learn about the early history of the home-brew movement in the United States (see "Organizations and Online Resources," page 247).

The Veggie Van

It would be difficult to talk about the grassroots biodiesel movement in the United States without mentioning the Veggie Van and its creators, Joshua and Kaia Tickell. In 1995, while she was finishing the production work on *Fat of the Land,* Sarah Lewison received a phone call from Kaia Tickell, who asked many questions about the cross-country trip. "We talked to Kaia a lot," Lewison recalls. "We tried to be helpful and gave her our recipes and told her about Tom Reed and Agua Dass."[22] After gathering as much information as they could, the Tickells organized a similar cross-country venture of their own. In 1997 and 1998 they managed to create a highly effective media buzz about their adventure as they toured the United States in a 1986 Winnebago that had been repainted with bright, impressionistic sunflower designs. In tow behind the van was a crude biodiesel processor on wheels, dubbed "The Green Grease Machine," capable of making biodiesel from used cooking oils. Along the route of this 10,000-mile-plus odyssey, the Tickells stopped regularly at fast-food restaurants to refuel, and they took advantage of the publicity generated to promote biodiesel at every possible opportunity. (They also visited Dr. Reed in Colorado.)

After the trip ended, Joshua Tickell wrote his popular how-to-make-it-yourself book, *From the Fryer to the Fuel Tank: The Complete Guide to Using Vegetable Oil as an Alternative Fuel.* A second tour with the Veggie Van added more than 10,000 additional miles to the vehicle's odometer. Joshua Tickell continues to travel at home and abroad, lecturing extensively on the benefits of biodiesel, and in 2004 he was working on a feature documentary film, *Fields of Fuel,* as well as an updated book about his favorite subject.[23]

Biodiesel Online

Since the mid-1990s, small-scale biodiesel production has continued to expand across the country. But because home-brewers tend to be geographically scattered, communication and exchange of information was initially a problem and progress was slow. However, with the spread of the Internet and the proliferation of online resources—especially various biodiesel community forums—the physical barriers to the sharing of information have disappeared. Some of those early forums have evolved into highly popular online resources such as the Biodiesel Discussion Forum (biodiesel.infopop.cc) and Biodiesel*Now*.com (www.biodieselnow.com). They typically contain hundreds (or even thousands) of postings on a wide range of regional, national, and international biodiesel topics. Today, online biodiesel discussion has become a major part of the home-brewing community. "It's so exciting now that it's possible to have this kind of information so readily available," Nicole Cousino says. "It makes such a difference. When we first started, there was no easy way to disseminate and share biodiesel information. Now we can get information on how to make the fuel, as well as questions and answers and concerns and warnings. And it's all shared for free; I think that's really important."[24]

9

The Main Players

The U.S. biodiesel industry has come a long way from its humble beginnings in kitchen blenders and garbage cans. As part of that process, the number of dedicated biodiesel plants in the United States has grown from one in 1993 to twenty-five as of October 2004, with at least a dozen planned to open in the near future. Biodiesel production grew from a few hundred gallons in the 1980s to 500,000 gallons (1,754 metric tons) in 1999. Since then, production has multiplied exponentially, to at least 30 million gallons (105,240 tons) in 2004. While this figure is admittedly dwarfed by the German production of 650,000 tons in the same year, it still represents major progress in the development of the U.S. industry. Nevertheless, since around 35 *billion* gallons of diesel fuel are consumed for on-road use annually in the United States, it's clear that the biodiesel industry has a long way to go before it begins to have a significant impact on the nation's fuel sector.

The Process

As the U.S. biodiesel movement has matured, more sophisticated and efficient process technologies have been developed that have transformed production from a backyard science project into a full-fledged industrial-scale sector of the energy economy. In the early years, all biodiesel was made using the batch process. Increasingly, biodiesel is

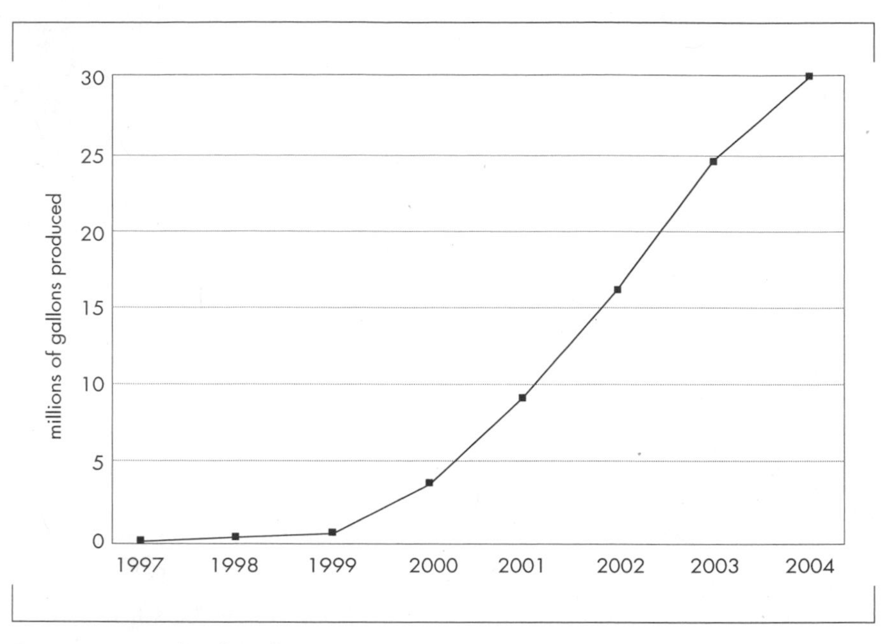

Figure 6. U.S. Biodiesel Production (from the National Biodiesel Board)

being produced by the continuous process, which was first developed specifically for biodiesel by Oelmuhle Leer Connemann in Germany in the early 1990s. Since that time, both processes have been fine-tuned somewhat by a number of technical developments, but the basic strategies remain the same. Although there are exceptions, biodiesel plants that produce more than 10 million gallons of biodiesel per year generally use a continuous-process design, while plants that produce less than 10 million gallons typically use the batch process.

"The batch process tends to lend itself better to smaller operations because it can be run as a single-shift operation very easily," explains Jon Van Gerpen, a professor of mechanical engineering at Iowa State University who has been involved in biodiesel research since the early 1990s, "whereas with continuous flow you pretty much are committed to round-the-clock operation, and most small plants are not prepared to do that."[1]

Regardless of the process, there are a number of different strategies for treating the feedstock triglycerides, according to Van Gerpen. "Some people start the process with the triglycerides and make biodiesel directly from them, while others first convert the triglycerides to free fatty acids and then convert them to biodiesel, although there are not too many people in this country who are using that second strategy," he says. But Van Gerpen acknowledges that the second approach is useful when the feedstock is already high in free fatty acids, as may be the case for animal fat or used cooking oil.

While there is a good deal of attention being focused on new advances in biodiesel process technology, Van Gerpen believes that it tends to be somewhat misplaced. "There's a lot of new technology being promoted, but the real issue for biodiesel is that the process doesn't cost that much compared to the total cost of production," he maintains. "Process is just not where the largest cost is, and coming up with a better process is not going to significantly affect the price of the product. My feeling, generally, is that process changes are not going to have much impact on commercialization."

So what will have an impact? Feedstocks and government policy, according to Van Gerpen. "The thing that is going to have the most dramatic impact on the market is what the government decides to do in terms of subsidies," he says. "And secondarily, the feedstock issue—whether there are any new feedstocks on the horizon such as mustard seed or something else that has the potential to increase the amount of oil that is available." Under current market conditions, most biodiesel feedstocks—especially virgin oils—are simply too expensive for biodiesel to be price-competitive with petrodiesel, according to Van Gerpen. While price is not everything, it certainly is the main issue that most consumers tend to look at first. Nevertheless, despite the general price disadvantage suffered by the industry relative to petrodiesel, most biodiesel advocates remain optimistic about future prospects—especially as petroleum prices continue to rise—and more and more biodiesel production facilities have been coming online in recent years.

The average-size commercial biodiesel facility in the United States has an annual production capacity of about 2 million gallons (that figure will soon grow as the size of proposed plants increases). Although construction costs vary somewhat, a ballpark figure for a new biodiesel plant is around $1 per gallon of annual production capacity. Actual biodiesel production costs generally run somewhere between $1.50 and $2.40 per gallon, with the cost of feedstock making up about 70 percent of those production costs. At about 20 cents a pound, soybean oil has traditionally been the lowest-cost virgin oil in the United States. It takes about 7.3 pounds of oil to make 1 gallon of biodiesel, bringing the feedstock cost to roughly $1.50 per gallon. (Soybean oil has risen to about 32 cents per pound recently, increasing the feedstock cost to around $2.34 per gallon and causing many biodiesel producers a lot of headaches.) Biodiesel begins to be price-competitive with petrodiesel when the feedstock costs for biodiesel are below 10 cents per pound, clearly demonstrating just how difficult it is for biodiesel to compete on the open market. The only common feedstock that is near this price threshold is "yellow grease" (recycled cooking oils), at about 12 cents per pound. Approximately half of the existing biodiesel plants in the United States are now able to switch feedstocks from virgin oils to recycled oils or vice versa without having to make changes to the hardware system in their production facilities.[2] This is a trend that undoubtedly will accelerate as the industry continues to look for more effective ways to level out the ups and downs in feedstock prices. Rendered animal fats are another low-cost feedstock, and the supply may increase due to concerns over bovine spongiform encephalopathy (mad cow disease), but how (or if) this will affect the U.S. biodiesel market is currently a matter of some debate within the industry.

Large U.S. Producers

Here is a listing of currently active plants in the United States (as of October 2004), arranged by state, according to the National Biodiesel Board:

- California has five plants, including American Bio-Fuels LLC, Bakersfield; Biodiesel of Las Vegas, San Luis Obispo; Bio-Energy Systems LLC, Vallejo; Imperial Western Products, Coachella; and Procter & Gamble, Sacramento.
- Texas boasts five plants: Corsicana Technologies Inc., Corsicana; Huish Detergents, Pasadena; Sun Cotton Biofuels, Roaring Springs; Texas Envirofuels, Poteet; and Texoga Technologies, Oak Ridge.
- Iowa is home to three plants, including Ag Environmental

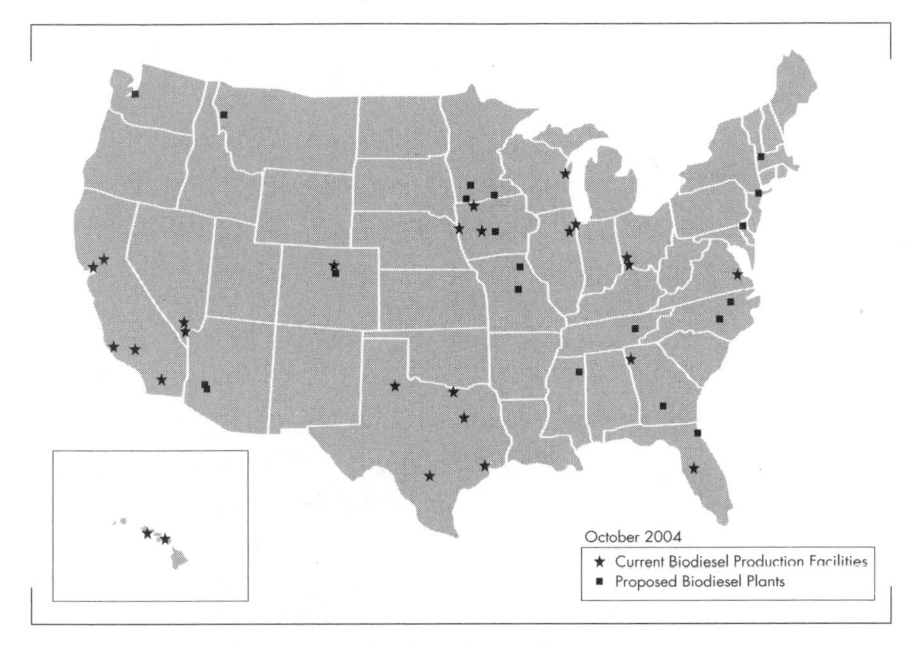

Figure 7. Current and Potential Biodiesel Production Plants (from the National Biodiesel Board)

Products, Sergeant Bluff; Soy Solutions, Milford; and West Central Soy, Ralston.

- Hawaii has two plants owned by Pacific Biodiesel, one at Honolulu and the other at Kahului.
- Illinois has two plants: Columbus Foods, Chicago; and Stepan Company, Millsdale.
- Colorado has one plant: Rocky Mountain Biodiesel Industries, Berthoud.
- Florida has one plant: Purada Processing LLC, Lakeland.
- Georgia has one plant: U.S. Biofuels Inc., Rome.
- Kentucky has one plant: Griffin Industries, Cold Spring.
- Nevada has one plant: Biodiesel Industries, Las Vegas.
- Ohio has one plant: Peter Cremer (TRI-NI), Cincinnati.
- Virginia has one plant: Virginia Biodiesel Refinery, New Kent.
- Wisconsin has one plant: Renewable Alternatives, Howard.

Some of these production facilities have already been described in chapter 8. The remaining list is too long for individual profiles. Consequently, what follows is a sampling of a few companies that demonstrate varying approaches to production, feedstock, or marketing strategies.

Griffin Industries

During the initial flurry of biodiesel plant construction in 1996, a family-owned rendering company in the state of Kentucky became interested in the potential that biodiesel offered. Founded in 1942, Griffin Industries of Cold Spring, Kentucky (about eight miles south of Cincinnati, Ohio), is the nation's second-largest rendering company. Like other companies in the business, Griffin takes what most people don't want—inedible waste from food plants and used cooking oils—and turns these ingredients into animal feed, organic fertilizer, industrial oils and fats, and, in recent years, biodiesel.

After carefully studying the market for a number of years, Griffin Industries decided to take the plunge into biodiesel in 1998 with the construction of a 1.6-million-gallon-capacity plant. The Austrian firm BioDiesel International Anlagenbau GmbH (BDI) from Graz provided the multifeedstock batch-processing technology for the facility. "We've been involved in biodiesel longer than anyone else in the rendering industry," says Dennis Griffin, chairman of Griffin Industries and one of five sons of company founder John L. Griffin who are still active in the business. "We spent about two years in research going around the world looking for a process that would work with high-free-fatty-acid feedstocks, which are primarily produced in our industry. It took a while to find that technology, but we finally discovered it in Austria. They had been at it longer than we had, so we took advantage of that experience and hoped to eliminate some of the trial and error that they had started out with."[3]

The plant, which opened in February 1998, is a state-of-the-art facility that is viewed as a prototype for larger plants to come in the future. "Even though it's a small batch plant, it's highly automated," Griffin says. "That's what made it cost so much, but we wanted to put the full operating system to the test in a small plant for shaking out all of the issues that we thought might be there with a much larger facility." The plant has performed as designed, and the support from BDI has been good, according to Griffin. "As I recall, on day four after we started it up, we were making ASTM-quality [i.e., meeting the standard set by the American Society for Testing and Materials] product, which is almost unheard of on starting up a brand-new process," he says. The plant is so automated, in fact, that only two people need to be present while it is in operation. And the fact that even two people are there is mainly for safety reasons, rather than for workload reasons, according to Griffin.

Griffin produces its Bio G-3000 Premium Biodiesel brand at its Butler, Kentucky, plant for customers in about fifteen states, including underground mining companies and trucking firms, and even for trials in the river transport and railroad sectors. But Griffin's biggest market is school buses. "Our main customer is the school bus industry," Griffin says. "They are becoming the primary users because of the diesel exhaust issue with the kids." And Griffin sees considerable room for additional growth in this segment of the market. Much of Griffin's biodiesel sales are of a B20 blend. "We have quite a few B20 customers, and that dramatically increases in the nonattainment areas during the summer months," he continues. "But we're trying to get away from that market to try to keep our production facilities running twelve months a year, not just four or five months during the summer for the B20 business." In order to accomplish this, Griffin is actively promoting a wider range of biodiesel blends for a variety of different purposes.

One of the biggest challenges for Griffin has been the company's ongoing effort to get government regulators as well as other feedstock organizations to understand that "feedstock neutrality" is vital for the future success of the biodiesel industry. Put simply, this means that regulations (or regulators) should not erect unfair barriers against, or provide unequal incentives for, one particular biodiesel feedstock over another. Animal fat–based biodiesel has not received a fair share of support for many years, according to Griffin. "It's been a constant battle," Griffin acknowledges. "For a long time we were the only ones who had an interest in this or had a plan or any involvement, so it was kind of a solo battle for the first three or four years." Today, more and more people in the industry are coming to understand the importance of this issue and are actively working to level the playing field (for more on this issue, see chapter 10).

Despite these problems, Griffin Industries has plans to significantly expand its biodiesel activities. Engineering has been completed on two new, much larger, continuous-process, multifeedstock plants, and specific sites have already been selected for them. Four additional plants are in the engineering stage, with sites yet to be determined. "We see ourselves probably in about a five- to six-year period of having eight to ten biodiesel facilities, dependent upon some of the legislative issues we are still fighting in Congress currently," Griffin says.[4]

World Energy Alternatives LLC

World Energy Alternatives LLC holds a unique position in the U.S. biodiesel industry. It is the nation's leading *supplier* of biodiesel, but until fairly recently it did not own any production facilities. World Energy's story begins in Massachusetts in 1994, when businessman James Ricci and a group of other investors purchased an idle oleochemical plant located in Quincy from the Procter & Gamble Company, renaming their venture Twin Rivers Technology. Most biodiesel literature credits Twin Rivers with being one of the first nondedicated commercial producers of biodiesel in the United States, but this is simply not true. From the start, the new owners *intended* to gradually shift the plant to the production of biodiesel, but after about two years the company had made very little progress with its biodiesel initiative—and absolutely no biodiesel of its own. "In fact, they had actually gone backward," recalls Gene Gebolys, who was hired to manage the biodiesel operation in 1996. "They had become a pretty good fatty-acid manufacturer, which is what P&G had done there previously, but they were not having much luck in the biodiesel arena. We bought biodiesel from Italy, Germany, and a number of U.S. oleochemical companies. The thinking was that, regardless of who made it, the focus had to be on selling the product."[5] In its earliest days, however, Twin Rivers had mistakenly positioned their product as an alternative to other alternative fuels—especially compressed natural gas

(CNG)—rather than as an alternative to petroleum diesel. As a result, environmental groups in both Boston and New York, who supported the use of CNG in transit fleets, were actively campaigning against Twin Rivers' introduction of biodiesel.

Gebolys worked hard to rescue the troubled biodiesel division and, in late 1997 did succeed in lining up a number of promising contracts with the Boston Transit Authority and the New Jersey Transit Authority. But the environmental activists remained adamant in their opposition, and by early 1998 Gebolys came to the conclusion that something drastic had to be done. "The biggest obstacle was that nobody had any real idea about what it was that we were selling, or any understanding of the value of what we were offering," he says. "And we were handicapped right at the starting gate without the help of our natural allies in the environmental movement." Gebolys reluctantly recommended selling or closing the faltering biodiesel initiative in order to save the larger and more successful oleochemical operation. He then tried to find a buyer for the biodiesel division. When that plan failed to attract any serious interest, Gebolys decided to set up a new company of his own that was focused on energy rather than oleochemicals. "Despite all the problems, I thought it had promise, but not at Twin Rivers," he says. Gebolys spent some time looking for financial backing for his plan and eventually attracted the support of Gulf Oil. With several biodiesel contracts spun off from Twin Rivers in hand, Gebolys launched World Energy in 1998.

At first World Energy bought all of its biodiesel from Procter & Gamble in Cincinnati, Ohio, and Gebolys spent the next few years building a customer base and a market. In 2000 World Energy entered into an agreement with OceanAir Environmental of California to buy the entire output of the former NOPEC plant in Lakeland, Florida (which Ocean Air was about to purchase from NOPEC). Subsequently, World Energy also contracted with AGP in Sergeant Bluff, Iowa, and several other producers for a portion of their annual biodiesel production. Then, in the summer of 2002, OceanAir broke their contract with World Energy, according to Gebolys. Legal action soon followed, and the case was eventually resolved with a settlement in favor of World Energy, which took over operation of the 18-million-gallon-capacity, batch-process Lakeland plant in the fall of 2003. "It does seem to be the Bermuda Triangle of the biodiesel industry," Gebolys says of the troubled former OceanAir, former NOPEC plant. "But early in 2004 we made a significant investment in capital improvements to make the Lakeland plant a top-notch, modern facility," he says. World Energy, with offices in California, Texas, Florida, and Massachusetts, is now a marketer, a distributor, and— finally—a producer of biodiesel.

The fact that the company is involved in so many different aspects of the biodiesel industry gives World Energy a number of advantages in the marketplace, according to Gebolys. "We are the only nonregional player in the United States," he says. "We are able to pull our product from many plants, so we have stable supply relative to the rest of the industry, which tends to go up and down. Also, we cost-average between these multiple plants, so we tend to have very stable pricing relative to other companies, who tend to be subject to swings in feedstock costs and so on. We're just more stable in supply and price and also more reliable. If someone has a problem with their plant we just get our product from a different plant. By almost any measure, our participation has been a stabilizing force in the growth of the industry."

But there is another important reason for the company's general stability: feedstock flexibility. World Energy has the ability to supply biodiesel from both virgin and recycled feedstocks, which allows it to average out some of the worst price fluctuations in the commodity markets—especially the recent price spike in the cost of soybean oil. "The fact that prices have been all over the place recently has been somewhat problematic, but we're in the energy business, and price swings are not unusual," Gebolys notes. "In fact, there has been better price stability in biodiesel than in petroleum. So, in terms of feedstocks, we don't have a particular ax to grind one way or the other; we just want what's best for our customers. We don't say that virgin and recycled are the same, because they're not, but we also don't say that one is inherently bad while the other is inherently good, either."[6]

Imperial Western Products

Headquartered in Coachella, California, Imperial Western Products has been in business since 1966 as a diversified feed commodities business. These commodities include feeds, oils, soaps, methyl esters, and glycerin. Imperial Western has been developing its state-of-the-art, batch-process biodiesel production facility in Coachella since 1999. "We designed and built the system ourselves from the ground up," says Bob Clark, sales manager for Imperial Western's biodiesel division. "We got into biodiesel because we were already large processors of used restaurant cooking oil."[7]

The facility, with an annual capacity of 12 million gallons, went into commercial production in 2002. Other than the self-designed aspect, the biodiesel plant has fairly standard elements found in most batch facilities: a receiving area where the feedstock is unloaded, reactors where all of the ingredients of the recipe come together and interact, a separation tank where the biodiesel and glycerin settle out and go their separate ways, a wash tank where residual impurities are removed from the biodiesel, a centrifuge where the water is removed, and finally a flash evaporator, which eliminates the alcohol. The glycerin by-product of the reaction is processed and then consumed either locally or sold for export. When the plant (which makes up about one-third of the total oleochemical facility at Coachella) is running at full capacity it is staffed by about ten employees, according to Clark.

Imperial Western makes two different types of biodiesel under the Biotane brand name: Imperial Biotane, from recycled restaurant oil, and Supreme Biotane, from virgin vegetable oil. In early 2004, however, due mainly to the high cost of virgin soybean oil, the company was producing biodiesel only from used oil. "At the moment, our feedstock is mainly recycled cooking oil," Clark says, "but we can use any other virgin seed oils with no problem. That's the beauty of the batch system; you can switch very easily from one feedstock to another." Imperial Western's marketing area is "everything west of the Rockies," and the company's primary customers are governmental fleets, both state and federal, including the military.

Although there are no immediate plans for further expansion of the biodiesel division, Clark says it might be a possibility at some point in the future "if things ever work out with federal energy legislation, or if the state of California becomes more supportive of biodiesel." The state's largest concern, according to Clark, is nitrous oxide (NOx) emissions associated with the combustion of biodiesel in diesel engines. "But with recycled cooking oil as a feedstock, you really don't have any significant NOx increase, particularly at a B20 blend level," he maintains. Another problem with certain state agencies and some members of the general public, according to Clark, is a lingering aversion to anything that even mentions diesel. "We did ourselves a great disservice by ever calling biodiesel 'bio*diesel*'—we should have called it something else," he says. "The word *diesel* just has an unfair bad association in the minds of some people. The perception with the general public is negative, and it shouldn't be." Another problem with some state officials, according to Clark, is that they oppose anything that prolongs the use of fossil fuels. "And biodiesel, in its most widely used form, depends on a mutual support structure with the petroleum industry," he says.[8]

American Biofuels LLC

In recent years there has been a trend toward the use of standardized, modular production units in new biodiesel plants in the United States. One company that has developed this strategy is American Biofuels LLC (ABF). In November 2002 ABF signed a joint-venture agreement with Hondo Inc. (an organic fertilizer manufacturer) to build a 35-million-gallon-per-year biodiesel plant at Hondo's existing 36-acre Bakersfield, California, facility. The new plant uses ABF's proprietary continuous-flow reactor/separator module, which is designed to produce 2.5 million gallons per year per module. Additional modules can be added quickly to increase plant production capacity as the biodiesel market expands. In November 2003 the new plant completed its first week of production test runs from its first module with various percentages of virgin and recycled cooking oils, and a short time later the first fuel produced by the facility was delivered to Hondo Inc. for use in its off-road equipment.[9]

In December 2003 a second 2.5-million-gallon reactor unit was delivered to the Bakersfield site, and after a brief shutdown for its installation, the expanded facility, with a combined annual capacity of 5 million gallons, started up again. ABF has registered with the U.S. Environmental Protection Agency (EPA) three different biofuel blends that will be sold under the Green Star trademark label (Green Star Products Inc. owns a 35 percent equity interest in ABF). In January 2004 ABF and Green Star assumed 100 percent ownership of the Bakersfield biodiesel plant when they bought out Hondo's 50 percent ownership of the joint venture.

ABF has announced plans to build a new biodiesel plant within the Greater Phoenix area. The site's infrastructure is large enough to allow for the processing and production of 10 to 15 million gallons of biodiesel per year. The new facility is expected to be ready for initial operation about three to four months after the start of construction. The modular plant design and the quick assembly techniques used at the Bakersfield plant promote this remarkably short blueprint-to-production time.[10] First there was Dolly the cloned sheep. Now there are cloned biodiesel plants.

Best BioFuels LLC

Although it's not on the list of currently active producers mentioned earlier, there is a new biodiesel project in Texas that is unique for several reasons. Best BioFuels LLC of Austin, Texas, is constructing a 12-million-gallon-annual-capacity, multifeedstock, continuous-process biodiesel plant in Fort Worth that will use biomethanol produced from swine waste in Utah as part of the biodiesel manufacturing process. "There is no other commercial biomethanol production on this scale in North America that I am aware of," says Jerrel Branson, president of Best BioFuels. The use of the bio-based methanol adds a renewable ingredient to the biodiesel recipe and is a prime example of an increasingly popular concept called "waste exchange," in which the waste from one commercial process is used as the input for another. Best Biofuels also owns the $20 million Utah biomethanol facility, which began production in May 2004.[11] Smithfield Foods of Smithfield, Virginia, is a majority owner of Best BioFuels.

The company has had a pilot biodiesel plant operating in Houston since early 2004. The plant, which has a production capacity of 1.4 million gallons per year, has been using brown grease from rendering plant sludge as a feedstock. At between five and ten cents per pound ($0.40 to $0.80 per gallon of biodiesel produced), brown grease is the cheapest feedstock currently available. The cost advantage of using brown grease is dramatic when compared with the $0.32 (current price) per pound for soybean oil ($2.34 per gallon of biodiesel produced), especially since feedstock costs represent such a large percentage of total cost of production. "Up to now, number 2 brown grease has not been used as a feedstock," Branson says. "But we've got some technology in the front end of the plant that we developed ourselves that allows us to take the brown grease and reduce the free-fatty-acid content to levels approaching that of seed oils." The initial biodiesel produced from the grease has been well within the parameters of the ASTM standard and was lighter in color than raw soy oil, according to Branson. "So we're pretty excited about the implementation of this in our full-scale plant in Fort Worth," he says. "We hope to have it up and running in March 2005."

"The biggest challenge was to make sure we had a system that was forgiving enough to accommodate variations in fats without having to go through the exercise of having to adjust the process for each load," Branson continues. "We can take poultry fat or pig fat or cow fat, and if there was enough lamb fat we could do that too. The challenge has been to have the flexibility without increasing the capital costs and spending a lot of time adjusting the system." In addition to being flexible, the in-house-designed process system is going to be significantly faster than most other systems, according to Branson. "We're very excited about it," he says about the new biodiesel operation. "I think we will be able to sell all that we can produce, particularly with the apparent new higher plateau for petroleum diesel pricing and with our lower feedstock costs. Our intent is to be able to sell a blend of B20 biodiesel at petroleum parity, including taxes, and the economics of the facility that we are seeing now is bearing that out."[12] Selling biodiesel at the same price as petrodiesel is every biodiesel producer's dream. Best BioFuels appears to be on the verge of achieving that significant milestone.

Blue Sun Biodiesel

Blue Sun Biodiesel of Fort Collins, Colorado, is an unusual, vertically integrated agriculture energy company that also deserves mention. Founded in 2001, Blue Sun's objective is to bring the cost of biodiesel in line with that of petroleum-based diesel fuel. In order to accomplish this goal, the company has embarked on an interesting series of market-based initiatives that begin with the customer and then work backward to the feedstock—the exact opposite of the strategy that has been followed by much of the U.S. biodiesel industry.

"What we really need to do is focus on the market, and that is what Blue Sun has done," says Jeff Probst, the company's president and CEO. "If we rely on the government sector and supply-side players like soybean farmers it's not going to get done, in my opinion. So what we're doing is generating demand in the marketplace based on our customers' needs and then pulling the product through the distribution network from the feedstock suppliers. That, basically, tells us how to build a viable business and a viable industry. And if you don't start in the marketplace, you're going to build something that isn't going to fit—you end up with the proverbial square peg in a round hole."[13] Probst, who previously worked for many years in the new product and technology division at Duracell, says that he and his colleagues always started with manufacturers to find out what the companies needed for their devices and then worked backward to develop their batteries. "That was a winning strategy at Duracell, it's going to be a winner here at Blue Sun, and it's going to be a winner for biodiesel too," he declares.

In order to make its market-driven model work, Blue Sun has created a network of partners, each of whom has a specific role to play in the company's larger plan.

Blue Sun is working with regional farmers through several new farmers' cooperatives to reduce the cost of biodiesel fuel through the development and production of low-cost oilseed crops for dryland agriculture on the high plains of Colorado, Kansas, and Nebraska. Initial trials have focused on mustard, but other crops such as false flax and rapeseed varieties are being investigated as well. Blue Sun has received grants from the Department of Energy (DOE) and the U.S. Department of Agriculture (USDA) to develop these new crop varieties that are best suited for high oil yields. The company has also set up what is essentially a franchise system of blenders, dealers, and retailers who move the fuel from the producer to the customers. Some of these customers include the City of Boulder, the University of Colorado, Aspen Homes of Colorado, and the Rocky Mountain National Park. To complete the picture, Blue Sun is also building a 3-million-gallon-capacity, continuous-process biodiesel production facility in Alamosa, Colorado, and hopes to have it in operation by the middle of 2005.

One of the main strategies Blue Sun is following is to encourage the production of industrial oilseed crops that will not impact the food supply. The idea is to grow biodiesel crops on unused cropland during the winter wheat crop rotation, providing farmers with additional income while not taking cropland out of food production. "Without taking away any winter wheat production, we can develop a $1 billion industry easily right here in Colorado, Nebraska, and part of western Kansas," Probst predicts. "And it could go up to $2 to $3 billion, and there would be positive impacts across the board from planting these new crops. We hope to set this up so 75 percent of that money would be going into our rural communities that are struggling right now, instead of sending it overseas to the Middle East." Probst admits that the Colorado market has been a challenge, but he is enthusiastic about his company's success so far. "We're getting a very good response from our substantial marketing and educational efforts," he reports. "I think people are smart and want to do what's right, as well as what's economical, and if you provide them with the opportunity they will participate."[14]

Small Producers

Although the market is increasingly dominated by larger and larger producers, a discussion of the biodiesel industry would not be complete without at least mentioning the role played by the small producer. These days, a small producer would probably fall somewhere in the range of between 100,000 and 500,000 gallons per year. Consequently, it's fairly safe to say that with a production capacity of between 15,000 and 30,000 gallons, John Hurley's Dog River Alternative Fuels Company in Berlin, Vermont, is one of the very smallest commercial biodiesel producers in the nation. Hurley, a logger and sawmill operator who manages 1,200 acres of Green Certified Forest land in Berlin, was concerned about global climate change and the impact his own diesel-powered equipment was having on water and air quality. After reading Joshua Tickell's book, *From the Fryer to the Fuel Tank*, in 1998, Hurley began to think about ways he could integrate biodiesel into his forestry and lumber operation on Chase Mountain.

In October 2000 the Vermont Sustainable Jobs Fund sponsored a conference, "Building an Ecological Economy." As part of that initiative, a challenge grant of $1,500 seed money was offered to someone who would put together a proposal to use the waste from any business to create a useful product. Hurley immediately thought about the possibility of making biodiesel from used cooking oil. "I thought, 'What the heck?' So I sent in an application," Hurley recalls. "Frankly, I didn't think I would get the grant." But he did. And that was the catalyst for Dog River Alternative Fuels. After discussing his plan with a number of other interested conference participants, Hurley and the group formed Dog River Alternative Fuels Company LLC the following April. "We became the first commercial biodiesel production company in the state of Vermont," Hurley says. The company's admittedly tiny pilot plant, housed in a small wooden shed behind Hurley's lumber-drying kilns, uses a standard batch process and on a good day can turn out perhaps 160 gallons, according to Hurley. "But we generally didn't make more than 120 gallons," he admits.[15]

Despite the modest output, Dog River managed to land an eighteen-month contract to supply biodiesel for a shuttle bus trial at the University of Vermont in Burlington. Dog River even has received a Governor's

Award for Environmental Excellence in Pollution Prevention for its work with biodiesel. But for a small producer like Dog River, joining the National Biodiesel Board or paying for ASTM testing is extremely difficult. This is the same dilemma faced by many other small producers across the nation. Dog River has been selling small quantities of biodiesel only for off-road (mostly farming and logging) vehicles and home heating. Still, Hurley is generally optimistic about the future of the biodiesel industry in Vermont and hopes in the near future to open a much larger 500,000-gallon-capacity plant in northern Vermont that will rely on used cooking oil as its primary feedstock. Once that source is fully utilized, Hurley hopes to tap Vermont oil feedstock crops grown on fallow land that is not used for food production. "It may require some creative efforts to do that," he acknowledges. But he's excited about trying.[16]

Cooperatives

There is a final variation on the small producer: the biodiesel co-op. By all accounts, there are quite a few co-ops of various types in existence (or springing up) around the country, but not too many of them have actually produced much biodiesel. However, they do seem to spend a lot of time and energy discussing, debating, and planning to make biodiesel. And because there are many different ways of achieving that end, there are many, many variations on the co-op theme. Nevertheless, biodiesel co-ops generally seem to fall into one of three main categories:

- very small-scale producer
- bulk buyer
- large volunteer producer group

The first category, very small-scale co-op producer, usually involves perhaps two experienced individuals who take care of all fuelmaking chores for themselves and a small group of other subscribers. This model, which offers relative safety, quality control, and ease of management, has been fairly successful—as long as it stays small.

The second category, the bulk-buy co-op, eliminates production chores completely (as well as possible production accidents) and normally offers less-expensive fuel to subscribers. Some bulk-buy co-ops use a large central storage tank where loads are delivered and where subscribers pick up their fuel. Other co-ops use distributed smaller storage tanks or barrels in multiple locations. Regardless of the storage strategy, a bulk-buy co-op still requires quite a lot of coordination and oversight to work smoothly.

Finally, there is the kind of co-op where everybody gets trained to make fuel on a volunteer basis. Of all the models, this one potentially presents the greatest challenges, according to people who have been involved with them. This is especially true for groups of more than five fuelmakers. The logistics involved with trying to train, maintain, and coordinate a large group of people to make consistently high-quality biodiesel safely are daunting. The Berkeley Biodiesel Collective in California (a producers' co-op), founded in September 2002, tried and tried to make this model work but ultimately failed. After a thorough reorganization of its efforts, the collective did eventually find a successful model that involved a major training initiative and a decentralized approach involving very small groups combined with a bulk-buyer's club for the purchase of commercial biodiesel.[17]

In New England, a new cooperative biodiesel venture is about to be launched, according to Tom Leue, president of Homestead Inc. of Williamsburg, Massachusetts. Co-opPlus, a member-owned energy cooperative in western Massachusetts, has been involved in a variety of renewable energy programs since the late 1990s. In April 2004 the 1,400-member organization voted to endorse a plan to build a 500,000-gallon biodiesel production facility in western Massachusetts based on a feasibility study provided by Leue (who has been making biodiesel in a very small commercial plant since 1998). Their plan is to rely solely on used cooking oil (yellow grease) from restaurants in the area surrounding the production facility as their feedstock. "Our model, which is rather unique, is based entirely on local resources and is intended to produce fuel for the local biodiesel market," Leue explains. "We find the business model to be economically very attractive on paper, and we think that we will be at least as profitable as some of the big producers. We hope this is true, and we're going work very hard to make it happen."[18] Assuming no major glitches, construction should begin in the fall of 2004, with production slated for the spring of 2005.

For a list of co-ops in the United States, see "Organizations and Online Resources" (pages 246–47).

A Maturing Industry

At the other end of the spectrum, along with the proliferation of new large-scale biodiesel production facilities, one of the most significant recent developments is that the big agricultural and petroleum oil companies in the United States, which have generally ignored the biodiesel movement for years, are beginning to take an active interest. In November 2003 Jerrel Branson, president of Best BioFuels, attended a board meeting of the National Biodiesel Board. "I looked around the room, and guess who showed up for the first time? ADM, the 800-pound gorilla of the agricultural oil business," he reports. But Archer Daniels Midland, an agribusiness giant and leading soybean processor, wasn't the only heavyweight in attendance. "Bunge Oil was there, Cargill, Tyson, Shell, and Ashland were all there too. The big boys are finally showing up, and it will become a big-boy business. This will no longer be a mix-it-in-a-garbage-can-to-fill-the-VW-microbus movement. Nor will it even be the folks designing a million-gallon-a-year plant. It's going to be the folks constructing large refineries and building real volume. Frankly, that's great for the industry, and that's the most significant change: the recognition that the industry is here, it's real, and we better be ready to do business."[19]

The National Biodiesel Board's executive director, Joe Jobe, agrees. "I feel good about this development, and I want more of it," Jobe says. "We are working in every way to educate the petroleum industry at all levels because the only way biodiesel will be truly successful is to integrate into the existing national liquid petroleum infrastructure. Without that key piece of the puzzle, biodiesel is just theoretical. It's pretty easy to see that the same infrastructure and vehicles that deliver diesel fuel can deliver biodiesel. But more important, biodiesel blends with petroleum diesel, and a blend strategy makes all the sense in the world. So I think this development is extremely important."[20]

Not everyone shares this enthusiasm for the large corporate model, but there is no doubt about it, the big players are finally getting involved. Shell Oil Company is already an associate member of the NBB, as are some very large regional petroleum marketers. Oil companies such as Gulf and BP have started marketing biodiesel. And Archer Daniels Midland—already the largest biodiesel producer in Europe—has been considering the construction of a biodiesel refinery in Minnesota and recently signed an agreement with Volkswagen to collaborate on the development of next-generation biodiesel fuels.

10

Biodiesel Politics

Although most Americans have been unaware of the biodiesel industry until fairly recently, those who have been involved for years understand that many different individuals and groups have played a part in the growth of the industry. Although in Europe the biodiesel movement developed essentially from the top down as a result of national government policy, biodiesel in the United States developed mostly from the bottom up, resulting in some uniquely "American" features in the U.S. biodiesel industry—and its politics. As mentioned previously, American soybean farmers have been the strongest and most persistent advocates of biodiesel for many years. Although there have undoubtedly been times when these farmers must have wondered if their efforts (and $35 million in financial support) would ever succeed, their persistence, accompanied by a lot of political clout in Washington, D.C., has finally begun to pay off. However, the initial dominance of soybean interests in the industry has also led to some rather bizarre political and practical dynamics that the industry continues to struggle with today, not the least of which is the simple fact that soybeans are not very high on the list of preferred oil-producing crops (see chapter 3).

But if you look a little closer at the industry, the internal and external political picture becomes even more complex. As the industry has rapidly expanded in recent years, the trend toward larger and larger production facilities (and the increasing participation of larger corporate players) has inevitably caused some of biodiesel's most ardent supporters in the "home-brew" and small-scale producer sector to feel increasingly threatened, marginalized, or even shut out. This is unfortunate, but not surprising. As a result, in the United States today there are two distinct segments of the biodiesel industry developing simultaneously. One follows the large agribusiness model, while the other pursues a small-scale, local model. Both groups produce biodiesel, and they are not necessarily mutually exclusive. Each has a productive and potentially complementary role to play as the industry evolves, but the tensions between them present the industry with one of its more difficult challenges.

Despite these internal difficulties, biodiesel has begun to attract a good deal of external political support in recent years from both sides of the aisle in the U.S. Congress. "That's one of our strongest selling points," says Neil Caskey of the American Soybean Association. "If you look at some of the policy successes that we've had on Capitol Hill, a lot of them are because biodiesel appeals to both political parties; it's not something that only Republicans or only Democrats are going to support. In light of the current energy situation we have, with increasing petroleum prices, if you have something that can address that issue, help farmers, and help the country all at the same time, it's an easy sell."[1] Increasingly bipartisan support makes the political case for the biodiesel industry easier to sell at the state level as well. It's hard for lawmakers, regardless of their political affiliation, to argue against helping farmers, creating new jobs, developing local economies, and strengthening energy security, to say nothing of numerous environmental advantages—all in one package.

Major Biodiesel Milestones

Although a series of important legislative and regulatory initiatives have helped propel biodiesel to its current status as the fastest-growing renewable fuel in the United States, a number of initiatives stand out as being major milestones: the ASTM standard, the Energy Policy Act, Environmental Protection Agency (EPA) approval, and several Presidential Executive Orders.

The ASTM Standard

While the biodiesel industry was still in its early developmental stage, the quality of the fuel was, at best, uneven. Trying to convince engine manufacturers to approve biodiesel use in their engines without some sort of common national standard was virtually impossible. In order to address this issue, in June 1994 a task force was formed in a subcommittee of the American Society for Testing and Materials (ASTM) to develop an ASTM standard for biodiesel. In June 1999 the preliminary standard was approved, and in December 2001 it was published in its final version, D-6751, for pure biodiesel (B100) in blends with petrodiesel up to 20 percent by volume (B20). Higher levels of biodiesel are allowed, but officially only after discussion with individual engine manufacturers (there are, however, plenty of individuals and an increasing number of fleets that use B100 all the time in their vehicles).

The publication of the ASTM standard was a major milestone in the development of the biodiesel industry, according to Charles Hatcher, the regulatory director for the National Biodiesel Board in Jefferson City, Missouri. "The importance for fleet managers was enormous; it assured them that the fuel had to meet a specification and that it was going to measure up to certain qualitative and quantitative measures," he says. The ASTM standard also helped to move biodiesel and the industry that produced it to a new level of legitimacy in the eyes of many people. In some respects, the new standard was "like a coming-out party for biodiesel," according to Hatcher.[2]

The Energy Policy Act

The Energy Policy Act of 1992 (EPAct) was passed by Congress to reduce U.S. dependence on imported petroleum by requiring certain fleets to acquire vehicles able to operate on nonpetroleum fuels. The act sparked interest in a wide range of alternatively fueled and powered vehicles and resulted in federal subsidies for research into electric vehicles, ethanol, methanol, and propane, as well as other alternative-energy technologies. In 1998 the EPAct was amended to allow fleets to meet a portion of their annual alternative-fuel-vehicle (AFV) requirements through the purchase and use of biodiesel in existing diesel vehicles. Previously, biodiesel and biodiesel-fueled vehicles had been excluded from the program. With this change, fleet managers could now comply with the EPAct by using an alternative fuel rather than by buying alternative-fuel vehicles.

The effect of the amendment on fleet usage of biodiesel was dramatic. In March 1999 three major fleets were known to be using B20 for EPAct compliance. By December of the same year, that number had increased to twenty-five and included the Ohio Department of Transportation, the U.S. Postal Service, General Services Administration, Alabama Power, and the U.S. Department of Agriculture. This represented a more than 700 percent increase in biodiesel users in just nine months.[3] On January 11, 2001, the final rule concerning the use of biodiesel to fulfill EPAct requirements was published in the Federal Register. This final rule amended Titles III and V of the Energy Policy Act of 1992, providing biodiesel-fuel-use credit to fleets that would otherwise have been required to purchase an alternative-fuel vehicle. Many fleets found this new rule to be a very attractive—though somewhat confusing—alternative.

"The biodiesel language that was added to the Energy Policy Act allowed biodiesel to be one of the means-of-compliance options," explains Charles Hatcher. "The phrasing of the legislation, however, has led to a lot of misunderstanding because the Energy Policy Act requires that a fleet—mainly government and utility fleets—achieve a certain number of 'credits.' For example, if your fleet was going to purchase ten new vehicles in a calendar year, it meant that you were going to need five credits or 50 percent of the ten vehicles to be in compliance under the provisions of the original act. Now, that figure has been raised to 75 percent, so in this example you would need 7.5 credits, which equates to seven and one-half vehicles." The biodiesel language allowed fleet operators to get one credit for the use of 450 gallons of B100. There were, however, a couple of provisos: the biodiesel had to be used in at least a B20 blend and it had to be used in medium-duty vehicles with a gross weight of at least 8,500 pounds. "If they did all of that, they could then get up to half of their required credits with biodiesel," Hatcher continues. "So, in this example, 3.75 vehicle credits could be achieved with biodiesel." The remaining credits would have to be achieved with a different fuel or AFV, the idea being to encourage reliance on a range of different fuels and vehicles.

If the ASTM standard was the coming-out party for biodiesel, then the amended Energy Policy Act of 1998 has really driven the industry ever since, according to Hatcher. "Those fleets and fleet managers are the ones who are purchasing the bulk of the fuel," he says.[4] The U.S. military alone accounts for at least 5 million gallons of biodiesel use and is the single largest buyer in the United States.

EPA Approval

Although the ASTM standard and the Energy Policy Act have been major milestones for the biodiesel industry, they certainly are not the only ones. In March 1998 biodiesel became the only alternative fuel in the nation to successfully complete the Environmental Protection Agency–required Tier I Health Effects testing under section 211(b) of the Clean Air Act. The testing clearly demonstrated biodiesel's significant reductions in virtually all regulated emissions (except for a slight increase in NOx). In May 2000 a Tier II subchronic inhalation testing study was completed as well. This testing demonstrated biodiesel's decreased threat to human health, especially when compared to the use of petrodiesel fuel. The total cost of the Tier I and II tests totaled over $2.2 million and was funded by the National Biodiesel Board.

As important as these tests were, the EPA has taken an additional step that virtually everyone in the industry agrees is going to give biodiesel an enormous boost in the very near future. In January 2001 the EPA finalized a rule requiring that sulfur levels in all diesel fuel for on-road vehicles be reduced from 500 parts per million to only 15 parts per million (a dramatic 97 percent reduction) by 2006. While this is good news for emissions, the downside of this reduction is that the refining process used to reduce the sulfur content in diesel fuel also reduces the fuel's natural lubricity agents, potentially causing excessive engine wear. Biodiesel is uniquely positioned to address this thorny issue. By adding to petrodiesel enough biodiesel to make up just 1 percent of the fuel, lubricity increases up to 65 percent. "Biodiesel has the sanction of the fuel injection pump manufacturers," says Daryl Reece, vice president for engineering, Pacific Biodiesel Inc. "They're okay with using 2 percent biodiesel to bring the lubricity level of diesel back up when the sulfur is removed. Biodiesel will be a great additive to accomplish this, so these injection systems can continue to operate properly. I think we'll see a big surge of biodiesel use in 2006."[5] Many others in the industry agree.

On May 11, 2004, the EPA took the sulfur reduction initiative one step further when it announced new regulations aimed at cutting the amount of smog-causing chemicals and particulates that come from off-road, diesel-powered vehicles and machinery such as ferry boats, harbor and river tugs, farm tractors, railroad locomotives, and heavy earth-moving equipment at construction sites. These vehicles account for about one-quarter of all the smog-causing nitrogen oxide and nearly half of the particulates from mobile sources, according to the EPA. First proposed a year earlier, the new EPA regulation requires petroleum refiners to lower the amount of sulfur in diesel fuel for such engines to 500 parts per million by 2007 and to 15 parts per million by 2010. This change will allow manufacturers to build cleaner-burning engines, since the fuel will contain much less sulfur, which damages catalytic converters and other emissions-control devices. The cleaner fuel, combined with the new engine standards, should reduce smog-causing nitrogen oxide and particulates from off-road vehicles and equipment by more than 90 percent, according to the EPA.[6] These new regulations will add even more momentum to the greater use of biodiesel already generated by the previously mandated reductions in sulfur levels in petrodiesel for on-road vehicles.

Executive Orders

Presidential executive orders can be useful tools for a wide variety of national agendas. Several executive orders have had a substantial positive impact on the biodiesel industry. On August 12, 1999, President Clinton signed Executive Order 13134, which called for the increased use of farm products, including agriculturally based biodiesel, instead of fossil fuels. Implementation of the order would effectively triple the use of bioenergy and bio-based products by the year 2010. President Clinton followed that up in April 2000 with Executive Order 13149, which called for a 20 percent cut in petroleum use by federal fleets. The continued strong increase in the use of biodiesel in federal fleets has unquestionably been encouraged by these two orders.

The Energy Bill

The fossil fuel and nuclear industries love it. Most environmental organizations hate it. Some environmentalists who are also biodiesel advocates reluctantly support it. Most major biodiesel producers, as well as Midwestern corn and soybean farmers, strongly favor its passage. "It" is the highly controversial, hotly debated, and long-delayed federal Energy Bill of 2002. Described as "the environment's worst nightmare" by some environmental groups, the Energy Bill had its roots in the now famous (or infamous) energy plan that resulted from the closed-door meetings between Vice President Dick Cheney and a group of energy industry officials shortly after the Bush Administration took over the White House in 2001. Containing huge subsidies for the coal, oil, natural gas, and nuclear industries, the Energy Bill also contains a few tidbits for renewable energy. One of those tidbits is a tax credit for biodiesel.

Republican Senator Chuck Grassley from Iowa has been a staunch supporter of biodiesel. The centerpiece of his efforts in support of the industry was his work on legislation to create biodiesel tax incentives that would include an excise tax credit for biodiesel when blended with petrodiesel. Grassley and Democratic Senator Blanche Lambert Lincoln of Arkansas played a key role in getting the legislation included in the Energy Bill that the U.S. Senate approved in 2002. However, the differences between the House and Senate versions of the Energy Bill could not be worked out in conference committee before the 107th Congress adjourned.

In February 2003 the two senators reintroduced the biodiesel tax provisions for inclusion in the Energy Bill, which promptly got mired in the 108th Congress due to the addition of a number of even more controversial provisions. In late July, after a total stalemate in the Senate, the new bill was dropped in favor of the legislation that had been approved in 2002. Passed again in the House, the new (old) Energy Bill became stalled in the Senate once again when a number of Republican lawmakers crossed party lines to help support a Democrat filibuster of the bill in late November 2003.[7] (A number of Democrats, mostly from Midwestern farm states, crossed party lines in the opposite direction to further complicate an already muddy picture of who was supporting what—and why—clearly demonstrating that there was no real consensus on the bill in either political party.)

In 2004, after some tinkering around the edges to try to gain support, the Energy Bill was reintroduced but remained hopelessly mired, with the fate of the biodiesel provision increasingly uncertain. The bill was again filibustered in the Senate, and on April 30 yet another cloture vote failed, leading many energy industry lobbyists to conclude that is was unlikely the bill would pass in an election year. With the Energy Bill effectively dead in the water, the biodiesel tax provisions were extracted from the bill by Senator Grassley and then added to a piece of tax legislation (S. 1637) in an attempt to get past the legislative logjam. The amended tax bill was passed by the Senate in May 2004 and moved to the House, where a similar piece of legislation that did not contain the biodiesel tax credit was also under consideration. If the House legislation passes, the action will then move to the Conference Committee, where differences between the two measures will be worked out, hopefully with the biodiesel tax provision intact.[8]

The tax credit amounts to one penny per percentage point of biodiesel blended with petrodiesel and would be available to excise tax payers and other fuel distributors who purchase biodiesel and blend it into diesel fuel, with the savings being passed on to consumers. Neil Caskey of the American Soybean Association says that his organization, which has played a key role in the passage of most important biodiesel legislation, has worked hard for passage of the Energy Bill. "I'm still cautiously optimistic that we will have a biodiesel tax incentive," he says. "Right now we're looking at between 20 and 25 million gallons of biodiesel produced in 2003. I've seen some estimates that if we get this tax incentive passed that figure could jump up to 40 or 50 million gallons, and it will continue to climb in the future as well. So we're hopeful that our greatest achievement is just around the corner."[9] As this book was going to press, the biodiesel tax credit contained in HR 4520, also known as the American JOBS Creation Act of 2004, was passed by Congress and subsequently signed into law by President Bush on October 22, 2004. The biodiesel industry's greatest achievement has finally arrived.

Subsidy Issues

In late 2000 the USDA's Commodity Credit Corporation (CCC) launched its U.S. Bioenergy Program. The program, which was intended to spur the construction of new biodiesel facilities, provides reimbursements to biodiesel and ethanol producers for converting a wide range of commodities into bioenergy. The program was also intended to provide a transition until biodiesel tax-incentive legislation was approved by Congress. As originally written, the biodiesel part of the program offered the subsidy only for soybean oil, which producers of biodiesel from other feedstocks widely condemned as being unfair. In the 2002 Farm Bill the program was reauthorized and the list of eligible feedstocks was broadened to include recycled fats, oils, and greases. But the revised subsidy structure still favored soy-based biodiesel with payments approximately twice as high as those for biodiesel produced from used cooking oil and animal fats. "That certainly has been a challenge," says Dennis Griffin, chairman of Griffin Industries in Cold Spring, Kentucky. "We never figured that the government would be our competitor, with these subsidies favoring soybean oil feedstock over the rendering industry's feedstock, even though we are a quality producer at a lower cost."[10]

The soybean portion of the CCC program is tied to the market price for soybeans. The price is set four times each year (ten business days before the beginning of each quarter). The quarterly subsidies are paid for new plant production on a production basis to those companies that have qualified as biodiesel producers (there is a $7.5 million annual cap per producer). The greatest payments are received in the first year, and company participation in the program ends after four years. The recent dramatic increase in soybean oil prices has caused the subsidy payments to soar from about $1.50 per gallon in the fall of 2003 to $2.50 per gallon for many producers in the second quarter of 2004. The price swings in feedstocks are somewhat dampened by the correlating price swings in the CCC subsidy program, giving producers a bit of a cushion to the ups and downs in the commodity markets.

The program payments to producers have generally been passed on to consumers and have reduced the retail price of biodiesel by $1.00 per gallon (or more in 2004). This price cut has been the single largest contributor to making market acceptance of biodiesel possible. It has also been largely responsible for the dramatic increase in U.S. biodiesel production (and plants) in recent years. "The CCC subsidies that came into play a few years ago have really helped to build the market," says Bill Ayres, formerly with Ag Environmental Products.[11] But not everyone is convinced of the program's long-term value. "The government subsidies are fairly lucrative right now," says Professor Jon Van Gerpen from Iowa State University in Ames. "But the current CCC program only encourages start-ups; it's not a good system to sustain an industry."[12] The program will come up for renewal in 2006, but its future is not clear. Some in the industry want to make further adjustments or improvements to it, while others, who view it as an impediment to an open market, would prefer to see it eliminated. "The CCC program is going to be a huge question mark when it comes up for renewal," admits NBB's Charles Hatcher.[13]

Feedstock Issues

The subject of feedstock neutrality is unquestionably one of the most contentious—and most important—issues facing the U.S. biodiesel industry today. But feedstock neutrality has not always been a high priority for the industry. The fact that the NBB was originally named the

National SoyDiesel Development Board is an undeniable indication of the organization's—and the industry's—roots. About five years ago, responding to criticism about a perceived soy bias, the NBB changed its bylaws to clearly state that the organization is "feedstock-neutral."

"No one feedstock will grow the market or be able to supply a fully implemented biodiesel market, so it really comes down to the need for cooperation at some point," Charles Hatcher says. "We simply can't take all of the soybean oil off the grocery store shelves, nor can we use all of the recycled cooking oil, or all of the animal fats, because there are other products and markets that are based on those commodities. In addition, cooperation is absolutely required so we can go through upturns and downturns in various commodity markets and still be able to provide a product." Despite the official neutrality of the NBB, it has been hard for some in the industry to swallow the lingering price differential in government programs between soy and other feedstocks. "Going back and telling your board that fifty cents on the dollar is feedstock-neutral has caused some disagreements," Hatcher notes.[14] What's more, when the Energy Policy Act was amended to include biodiesel in 1998, there was a federal guideline that limited fleets to purchasing only soy-based biodiesel. Producers using other feedstocks complained bitterly about being shut out of the market, and the limitation was recently relaxed to include biodiesel made from virtually any feedstock, as long as it met the ASTM standard. To its credit, the NBB was involved in the process to have the soy limitation removed from the federal purchasing guidelines.

Despite the fact that World Energy now owns the largest facility in the country specifically designed for utilizing yellow grease, Gene Gebolys, the company's president, is philosophical about the feedstock wars. "If we are not politically savvy enough to understand that if we don't have public support for our product, then we don't have an industry," he says, "and if we aren't savvy enough to understand that farmers have more political clout than renderers, then we are pretty foolish. Willie Nelson doesn't play his guitar for the renderers. It's important for us to understand that we absolutely need political support, and that the soy interests are in the best position to deliver that support. And we can't expect them to do that without getting their fair share of the return. After all, they are the ones who have been forking out the cash for this industry from the beginning and exerting their political influence. It seems naive to imagine that the soy interests should somehow step aside now." Having said that, Gebolys also understands that the soy interests need the cooperation and support of other feedstock producers to help stabilize the market. "Ultimately, the success of soy biodiesel is going to be dependent on the ability to get other feedstocks to dampen price swings and to help create a predictable market," he says. "You can't have an industry if people don't know whether they can get the feedstock or even roughly what it is going to cost them. That's ultimately where we are headed, a plurality of feedstocks."[15]

State Initiatives

In addition to national initiatives, the industry in general, and the NBB in particular, has been very active with a wide range of state-related programs, especially in the past few years, according to the NBB's Charles Hatcher, who now spends most of his time helping to get state biodiesel legislation passed. Essentially, most of these state initiatives are aimed at convincing state governments to try biodiesel. "Getting states to use biodiesel does so many wonderful things," Hatcher says. "First, the state is an immediate customer, so right there you've got sales. Second, it educates perhaps fifty thousand people at once, so it provides a lot of education quickly. You also get distribution because the state has vehicles not only in the capital but all over the state, so the fuel supplier has to start looking at other locations and retail stations."[16]

The proliferation of these state initiatives has begun to pick up momentum in the past few years and has led to several parallel and simultaneous developments. On the one hand are states that have had little or no biodiesel experience, and on the other hand are states that are becoming involved in increasingly complex biodiesel activities. Sometimes these two levels of experience are found within a single state.

"This is an industry that has grown up so quickly that we haven't had time to stop and congratulate ourselves," Hatcher says. "In the past twelve months I have gone from simply educating legislators about biodiesel basics—giving a very simple classroom discussion about what it is—to dealing with major tax issues. For example, someone who just started importing biodiesel from one state sells it to a distributor who is, in turn, exporting it to another state, and they want to know who owes the taxes, and to which state. Good question. Or a fuel dealer that just started selling biodiesel calls us the very next day about dyeing requirements, and none of us know what he is talking about. And then the next day we go from that to trying to understand the terminal rack system and the fuel distribution system—it's all moving very fast. I think we'll see much more growth on these two different levels, where we will still be trying to get people to understand and use biodiesel, but then also dealing with very complex trade-association industry issues."[17]

In many states lawmakers have recognized the importance of renewable fuels and have recently passed various types of biodiesel legislation. The pace of this activity has increased dramatically in the past few years. Approximately thirty-three biodiesel-specific pieces of legislation were introduced in 2003. Twenty-six of them passed. New laws approved recently in Minnesota and Illinois are good examples of this trend.

Minnesota

Although there have been a number of state biodiesel initiatives, Minnesota's groundbreaking "2 percent" legislation is unquestionably one of the most sweeping, and the first of its kind in the nation. On March 15, 2002, the state mandated that all diesel fuel in the state contain 2 percent biodiesel (B2). The implications of this seemingly insignificant percentage of biodiesel are enormous. Minnesota uses 831 million gallons of diesel fuel each year, and 2 percent of that equates to a new market demand for over 16 million gallons of biodiesel. The law will go into effect on June 30, 2005, on the condition that the state will have 8 million gallons of biodiesel production capacity by that date. "On behalf of the Minnesota Soybean Growers Association and all of the other farm groups and others who fought so hard for this bill, I can only say we are absolutely thrilled and grateful for the support of the legislature and the governor," said Ed Hegland, president of the association. "Biodiesel will prove that it is an exceptional fuel that will help rural economies, provide cleaner air, and allow Minnesota to lead the nation in insisting on the production and use of alternative, non-petroleum energy sources."[18] The new law has spurred a flurry of biodiesel initiatives, and a recent Minnesota Department of Agriculture study projected that on-road demand alone will generate a direct economic impact of over $212 million annually and create more than 1,120 jobs.[19]

Illinois

In June 2003 two important pieces of biodiesel legislation were signed into law by Illinois Governor Rod Blagojevich. One bill provided for a partial tax exemption on biodiesel blends, while a second bill offered grants of up to $15 million for the construction, modification, or retrofitting of renewable energy plants with a minimum production capacity of 30 million gallons of biodiesel a year. As a result, the Illinois Soybean Association, anticipating an increased demand for soybeans, estimated that the legislation would add five cents per bushel to the price of Illinois soybeans. With the state's annual production of soybeans at about 450 million bushels, the legislation was expected to boost the state's economy by more than $22.5 million.[20]

Small-Producer Issues

One of the most challenging political issues in the biodiesel industry is internal. As mentioned earlier, there has been a certain amount of ongoing tension between some small producers and some of the larger producers as well as with the National Biodiesel Board. Annual membership dues starting at $5,000 for the NBB were viewed by many small producers as deliberately exclusionary, and the NBB has been seen by some as simply a mouthpiece for large producers and soybean farmers. Aside from the high dues threshold, a related bone of contention has been the issue of access to the health effects data of the EPA Tier I and Tier II tests that the NBB paid for. Access to the data greatly facilitates biodiesel registration with the EPA. "What this boils down to is the issue of the NBB's core function," says Charles Hatcher. "Are we a benevolent organization trying to get people to do the right thing, or are we a business trade association? We sort of fill both of those roles, but we did pay for all of those tests, and we do have the registration. We aren't stopping anybody from going out and doing their own testing and getting their own registrations. Or they can use ours for a price. We set that price at $5,000. I know that some small producers say that's not fair. But on the other hand, I don't know of any business that can't afford a $5,000 licensing fee."[21] Fair or not, some small producers have become so frustrated with the situation that they have been exploring the possibility of forming their own industry organization that would be more in tune with their needs.

Such divisive issues are not helpful for the industry, and some of the participants have recently attempted to try and find a solution. At the National Biodiesel Conference and Expo held in Palm Springs, California, in February 2004, the NBB made an attempt to open a dialogue with small producers. "A lot of folks in California were interested in these issues, and I wanted to open a dialogue so we could find out just exactly what the problems were so we could try to address them," says NBB's Joe Jobe. "We had a lot of conversations, and I think it was incredibly positive. Based on those discussions, the governing board reactivated a small-producer working group to look into these issues. They'll be making recommendations to the full board and the governing board that will at least partially address these issues. I can guarantee you, however, that there will be folks on both sides that will not be entirely satisfied."[22] In July 2004 the NBB did, in fact, address the membership issue by deciding to create a new membership class, "small producer nonvoting associate member," for producers making less than 250,000 gallons per year, and also lowered the minimum annual dues from $5,000 to $2,500. As predicted, not everyone was entirely satisfied, but this move was an honest attempt to address some of these concerns.

One producer that has distanced itself from the NBB for many years is Pacific Biodiesel in Hawaii. But that situation is changing, according to Bob King, the company president. "We've had a long history of not participating in the NBB because of how they did business in the past, but I'm really excited about the new NBB," King says. "Joe Jobe and [NBB chairman] Bob Metz are great guys, and they are somehow able to work with this extremely diverse group of constituents and do some good things. That's just really great for the industry, and I'm really pleased with where the organization is at now. I have to give them a lot of credit for what they have done."[23]

Since King is located in Hawaii, he says that he has managed to stay out of the small producer/big producer squabbling on the mainland. But he still has some thoughts on the issue. "My opinion has always been that small production plants are going to be the industry-opening plants," he explains. "So, in an area that does not have any biodiesel, this is the size you start off with because it's economic; you can start making money with it right away. Then, after things catch on, the industry should move on to big plants. It's going to be a different group of people who are going to be doing those larger plants; it's a different mindset and a different production process, but I don't really have a problem with either group because they each have a role to play. As long as the company is doing a good job for its customers, the community, and the environment, then it doesn't matter if the company is big or small. I really hope that everybody gets over the infighting that's going on and realizes that we're all doing a good thing and there's plenty of room out there for everybody."[24]

But tensions between small and large producers are not unique to the United States, according to Dr. Rico Cruz of the Environmental Science and Technology Program of the Confederated Tribes of the Umatilla Indian Reservation in Washington State. Dr. Cruz, who is originally from the Philippines, is a pioneer in the production of biofuels and simple-process technology. Since the early 1990s he has been involved in a wide range of biodiesel projects in the United States and in countries such as India, Uganda, Bulgaria, Russia, Azerbaijan, and the Philippines. "Some of the bigger biodiesel producers are definitely trying to eliminate the smaller producers," Cruz maintains. "In many of the countries that I have visited the biodiesel producers do not have the support of their national governments. But once a large producer monopolizes the production, they try to get all of the government support. I've seen it happen, it's still happening, and I think that's a problem."[25] This is a matter of personal concern for Dr. Cruz, since most of the biodiesel projects he has helped set up are small-scale, community-based operations. Dr. Cruz is credited with bringing biodiesel to the Philippines as well as to the Nez Perce Indian Band in the Pacific Northwest.

The Ethanol Connection

The ethanol industry has had a significant impact on the development of the biodiesel industry in general, and on biodiesel politics in particular. The ethanol industry had a long head start on the biodiesel industry, with a well-organized and powerful lobby composed traditionally of Midwestern corn farmers, who have been able to attract and maintain substantial government support via their considerable political muscle in Congress. Up until the mid-1990s, if you were talking about a liquid biofuel in the United States you were basically talking about ethanol. Since then the industry has continued to prosper and grow. To put this in perspective, in 2003 the U.S. ethanol industry produced 2.8 billion gallons, while the U.S. biodiesel industry about 25 million gallons of biodiesel. This ten-to-one ratio is a clear measure of the political success of the ethanol industry, as well as an indication of the future potential for the biodiesel industry.

Nevertheless, the biodiesel industry has been able to use many of the same political strategies developed by the ethanol industry to gain government support, in part because some of the same people have been involved in both industries. And these people learned some lessons from their prior experience with ethanol. "Farmers who grow soybeans also grow corn, and a lot of them have been involved in the development of the ethanol industry," says Joe Jobe. "One of the lessons learned from the ethanol industry was that some of the technical issues on blending, storage, water absorption, and so on weren't totally worked out. So ethanol got a bad name and developed a bad reputation that to some extent is still following it around thirty years later. So the farmers who started biodiesel in the U.S. took a very long approach and wanted to get all of the technical, legal, and regulatory issues settled before they worked on commercialization."[26]

All of this experience and attention to detail has created a reasonably successful political record for the biodiesel industry at the national level. It is expected that this momentum will continue in the future. And what works at the national level can sometimes be transferred to the state level. Even in Hawaii—about as far from the Midwest as you can get—biodiesel producers have been able to make use of the momentum generated by the ethanol industry. "The ethanol industry has already done a lot of the groundwork for the biodiesel industry," Daryl Reece of Pacific Biodiesel says. "The state of Hawaii already had a tax credit for ethanol, so all we had to do was to piggyback biodiesel on the existing state tax credit for ethanol, and it was passed easily."[27] Similar strategies have worked in other states as well.

Straight Veg

Of all the political issues related to biodiesel, probably the most contentious is the struggle between the industry and individuals who promote the use of straight vegetable oil (SVO) as a diesel fuel substitute. This strategy, referred to by many enthusiasts as "straight veg," makes use of waste vegetable oil (yellow grease) collected from restaurants to fuel diesel engines *directly*, without first going through the transesterification process. This bit of alchemy is accomplished by adapting the vehicle to the fuel with a conversion kit (costing roughly $300 to $1,500 in the United States). Basically, the typical conversion kit involves adding a parallel fuel system that consists of a second fuel tank, a heater, an extra fuel line to the engine, a filter, and a control that allows the driver to switch back and forth between the two systems. The vehicle has to be started with diesel fuel from the first tank, and then, after the engine heats up and thoroughly warms the straight vegetable oil in the second tank, the driver manually switches to the second tank (some systems feature an automatic control for this). At this point the vehicle is running on straight vegetable oil. When the vehicle needs to be shut off, the process is reversed. The driver manually switches back to tank number one (containing the diesel fuel) for a few minutes to ensure that all the straight vegetable oil has been removed from the fuel line and engine.

While this strategy is relatively simple, the debate about it is anything but. Virtually all the big players in the biodiesel industry in the United States and Europe warn about possible engine damage and complain that straight veg users are breaking the law by not paying fuel taxes and giving the biodiesel industry a bad name when problems do occur. There is some justification for this view, since most of the general public is still not educated enough about various fuels to understand the finer points of these different strategies. On the other side, the straight veg camp (which also includes many backyard biodiesel producers) tends to have a conspiratorial mindset about the biodiesel industry's intentions and complains that the industry is spreading false information about straight veg. Unfortunately, the print and broadcast media cause additional confusion when articles or news reports sometimes confuse biodiesel and straight vegetable oil in their reporting.

The straight veg strategy appeals mainly to people who are mechanically inclined and who can remember to switch between fuel tanks at the appropriate moments (this is definitely *not* a good strategy for the absent-minded). In addition, there may be some potential long-term problems, especially if the straight veg is used carelessly. And since there is a limited supply of used cooking oil, straight veg is always going to be a somewhat limited option. Nevertheless, the lure of essentially free fuel is compelling.

Ironically, all the players involved in this controversy have the same general goal in mind: a shift from a fossil fuel to a renewable biofuel. But they follow different paths to reach that common goal.

11

Recent Developments

In the summer of 2002, as she performed around the country with Lyle Lovett, folk/rock/blues singer Bonnie Raitt was doing more than entertaining her many fans. She was the first U.S. entertainer to adopt biodiesel for use on tour. On what was dubbed the Green Highway Tour, the nine-time Grammy Award–winning artist traveled part of the route in a biodiesel-powered bus, visiting forty-two cities along the way and performing at major venues while raising awareness about alternative fuels.

"It's no accident that we're in danger of losing both our ecological and our economic well-being at the same time," explains Raitt, who created the Green Highway concept along with colleagues Kathy Kane and Harvey Wasserman. "I feel too many government and corporate policies are inseparably shortsighted, and we've created Green Highway to demonstrate that working in harmony with nature can offer real solutions for preserving both our planet and our prosperity."[1]

The tour was powered by a combination of biodiesel, solar energy, wind power, and hybrid-vehicle technologies. At each stop, the tour set up exhibits on alternative energy and handed out information to concertgoers.

Raitt is as well known for her long-time commitment to social activism as she is for her music. She has been involved with the environmental movement since the mid-1970s and was a founding member of MUSE (Musicians United for Safe Energy), which produced the enormously popular 1979 NO NUKES concerts at Madison Square Garden. Raitt has been especially active in the fight to preserve ancient forests, performing numerous benefit concerts, lobbying in Washington, D.C., and twice committing civil disobedience in support of more ecologically sound government and corporate policies.

But Bonnie Raitt isn't the only celebrity to promote biodiesel. In 2003 the popular folk duo Indigo Girls used biodiesel as part of a tour focusing on alternative energy and Native American environmental issues. The two-week Honor the Earth tour began April 10 in Northfield, Minnesota, and wrapped up on April 22, Earth Day, in Greeley, Colorado. Grammy Award–winners Emily Saliers and Amy Ray performed acoustic hits during the tour, but much of their performance included commentary on the connections between the environment, energy, Native American issues, and U.S. foreign and domestic policy. Proceeds from the tour went to Honor the Earth's Energy Justice Initiative to support alternative energy developments in Native American communities. "There is a strong connection between energy policy and environmental issues," said Indigo Girls performer Emily Saliers. "Using an alternative fuel like biodiesel is a way for us to be proactive in using less foreign oil. That's something we feel very strongly about."[2]

Also in 2003, the legendary modern rock music festival Lollapalooza used B100 to fuel the generators that provided power for its second stage and tent exhibits. Founded in 1991 by Perry Farrell, lead singer of the band Jane's Addiction, Lollapalooza was the largest U.S. music tour to use biodiesel to date.

In February 2004 veteran Canadian rocker Neil Young found a new way to support farmers by fueling the trucks and buses on his Greendale tour with B20. "Our Greendale tour is now ozone-friendly," Young said. "I plan to continue to use this . . . fuel exclusively from now on to prove that it is possible to deliver the goods anywhere in North America without using foreign oil, while being environmentally responsible." Young has supported American farmers for more than eighteen years through his participation in the popular Farm Aid benefit concerts.[3]

Even actress Daryl Hannah is a strong advocate for biodiesel. Hannah, who starred in memorable films such as *Splash*, *Blade Runner*, and *Steel Magnolias*, has been a long-time environmentalist and biodiesel activist.

She has educated many thousands of people about biodiesel through her appearances on television shows such as *The Tonight Show* and *The O'Reilly Factor,* as well as through numerous magazine interviews. Hannah was honored at the 2004 Biodiesel Conference & Expo in Palm Springs, California, for her volunteer advocacy of biodiesel as well as long-time use of the fuel in her own personal vehicle. "We have the technology to reduce greenhouse gases and grow much of our own fuel," Hannah said. "We have the technology to make sure no kid goes to school breathing dangerous toxins. We have the technology now, and that technology is biodiesel."[4]

Biodiesel Use Expanding

Although celebrities who use and endorse biodiesel attract a lot of media attention and are unquestionably helpful in raising public awareness, there are a lot of other people all across the United States who have gotten the message about the fuel's many advantages as well. Even former President Jimmy Carter has been an active proponent of biodiesel, working in his home state of Georgia to bring together industry representatives and legislative and government leaders to discuss making biodiesel production a reality there.

The pace of first-time biodiesel use has accelerated in the past few years to the point where news of yet another school district, city bus system, or municipality somewhere in the nation switching to B5 or B20—even B100—has become almost a weekly occurrence. It would be impossible to list them all here, but what follows is a brief summary of recent biodiesel news and events that will present a reasonably good cross-section of current U.S. biodiesel activity.

New Biodiesel Pumps Open

For quite a few years it has been possible to purchase biodiesel—if you knew where to find it—from a small producer or farmers' co-op. But for all practical purposes biodiesel was not readily available to the general public. As of 2001, that began to change. Sparks, Nevada (near Reno), appears to deserve the credit for having the first public biodiesel pump in the nation—but just barely. Officially opened on May 22, 2001, by representatives from Western Energetix and the Nevada Energy Office, the Western Energetix Cardlock location at 655 South Stanford Way in Sparks was the first public filling station to distribute biodiesel made from recycled cooking oil to the general public. The used cooking oil came from local casino resorts and restaurants and was made into fuel by Biodiesel Industries of Las Vegas with the assistance of the Nevada Energy Office and a grant from the U.S. Department of Energy. "The support we have received from the State of Nevada in getting this project from a dream to reality has been tremendous," said Russ Teall, president of Biodiesel Industries. "I hope this can show communities all across America that biodiesel can be made and used almost anywhere. When government and industry work together to solve problems, it's amazing what can be accomplished in a relatively short time."[5]

Less than twenty-four hours later, America's second public biodiesel pump was officially opened in San Francisco, California. The ribbon-cutting ceremony took place at the Third Street facility of Olympian Inc., with local, state, and company officials in attendance. World Energy Alternatives of Chelsea, Massachusetts, supplied the fuel. "As an oil company operating in today's economy, we believe offering biodiesel not only improves our corporate image, but also increases our competitive advantage," explained Tom Burke, Olympian's Division Manager of Cardlock and Mobile Fueling. "When renewable fuel becomes the rule and not the exception, we will already be a recognized provider."[6] Many other forward-looking fuel companies are beginning to have similar ideas.

Since these initial openings, public biodiesel pumps have been sprouting up all across the nation like mushrooms, from California to Connecticut and from Maine to Missouri—and a lot of states in between. As of late 2004, there were slightly more than three hundred pumps nationwide (compared with the more than two hundred thousand that offer gasoline and petrodiesel). The largest number, however, are still clustered in the Midwest. The National Biodiesel Board offers a map showing U.S. public biodiesel pump locations by state at www.biodiesel.org/buyingbiodiesel/guide. Scroll down to the "Retailers" section and click on **National Map of Retail Fueling Sites** and then individual states for details.

Biodiesel in Schools

There are many niche markets for biodiesel, but school buses, in particular, make a lot of sense. The fact that there are about 460,000 school buses in the United States—nearly six times as many buses as all the nation's public transit buses combined—is reason enough. But the fact that children—especially young children—tend to be more susceptible than adults to the toxic and potentially cancer-causing emissions from petrodiesel has been an even more compelling reason for school boards and parents across the nation to insist on school buses being switched to biodiesel fuel. The federal government has been helpful in this process. Congress included $5 million in the Environmental Protection Agency's budget for Clean School Bus USA, a cost-shared grant program designed to help school districts in cleaning up their bus fleets. The fact that the EPA received more than 120 applications requesting almost $60 million is a clear indication of just how popular the program has become.[7]

In 1997 the Medford, New Jersey, school district was the only one in the nation to run its fleet on biodiesel. But today thousands of school buses use the fuel. The Clark County, Nevada, school district now powers more than twelve hundred of its buses with biodiesel, making it the largest school bus fleet in the nation (and possibly the world) to use biodiesel. In Indiana more than a dozen school districts have switched fuels, while in the state of Kentucky nine school systems are now running six hundred buses on biodiesel. A total of about fifty districts nationwide have made the switch to biodiesel, and the numbers are growing almost daily. Two large Colorado school districts, Littleton Public Schools and Jefferson County Public Schools, have been using biodiesel successfully for several years, and Littleton has reported decreased maintenance costs since making the change.

Although the details are still a little sketchy, there is increasing evidence that other school bus fleets are also saving money by using biodiesel, even though the fuel costs more than petrodiesel. How is this possible? The savings are in reduced maintenance costs and increased mileage per gallon. Some of the strongest evidence comes from the Saint Johns Public Schools in Michigan, where careful maintenance records have been kept from both before and after biodiesel was adopted in April 2002. Saint Johns was the first Michigan school district to switch its entire fleet of buses (totaling thirty-one) to B20. The main cost savings have been due to extended intervals between oil changes, according to Wayne Hettler, garage foreman and head mechanic for Saint Johns. "I'm convinced," he says, "that we are able to extend the oil changes because the B20 burns cleaner and isn't dirtying the oil as quickly. We're using oil analysis to determine oil change times. We solely credit biodiesel for cleaning up the oil, thus saving the district the costs of oil, filters, labor, and the like. I challenge other fleets to 'read' their fleet records and make these cost-saving changes after switching to B20." Longer fuel-pump life due to biodiesel's higher lubricity and increased miles-per-gallon rating are also cited by Hettler as adding even more savings. "Pre-April 2002, our fleet's mileage averaged 8.1 miles per gallon. Now we average 8.8. That's a *huge* difference in MPG for buses," said Hettler. A combined savings of $3,500, even after the extra cost of the biodiesel is deducted, is predicted by the district for the two-year period.[8] If savings can be realized by this school bus fleet, it seems reasonable to assume that other fleets can do the same.[9]

On the East Coast, the Warwick, Rhode Island, school district not only uses biodiesel in its entire seventy-bus fleet but has been successfully heating three of its school buildings with biodiesel since 2001 (see chapter 4). But the school district has gone even further by integrating biodiesel into its classroom curriculum. Their program is modeled after the high-school curriculum on alternative fuels developed by the Northeast Sustainable Energy Association called "Cars of Tomorrow and the American Community."[10]

More Mass Transit Use

Transit buses use a lot of fuel, consuming on average about 10,000 gallons every year. Of the more than seventy-six thousand active transit buses in the United States, about 90 percent are still powered by diesel fuel; roughly 7.5 percent are running on compressed natural gas (CNG); 1.5 percent run on liquified natural gas (LNG); and a small number are electric, hybrid electric, or fuel-cell-powered (some of the hybrid electrics are diesel hybrid electric). Despite the increasing numbers of natural-gas-powered buses in recent years and the fact that orders for new diesel buses have gradually been declining, the size of the present diesel-powered fleet remains relatively steady.[11] There is, however, some anecdotal evidence that the boom in CNG buses may be slowing due to their higher cost, the great expense of support infrastructure, and continued price increases for natural gas. In any case, diesel-powered transit buses offer a substantial market opportunity for biodiesel. Among the many city bus fleets in the United States currently using biodiesel are those in Cedar Rapids, Iowa; Cincinnati, Ohio; Saint Louis, Missouri; Oklahoma City, Oklahoma; Olympia and Seattle, Washington; Raleigh, North Carolina; and Springfield, Illinois.

In 1993 Five Seasons Transportation & Parking (FST&P) in Cedar Rapids became one of the first bus fleets in the United States to use biodiesel, but in 1996 the program was dropped due to high fuel costs. Since then the cost of biodiesel has come down while the interest in renewable fuels has gone up. In March 2001 the program was revived and sixty buses began burning B20 again. Five Seasons was Iowa's first mass transit system to convert its entire diesel-powered fleet to biodiesel. Ag Environmental Products of Lenexa, Kansas, supplied the fuel. Using a combination of emissions control technologies and biodiesel, Five Seasons twenty-five-year-old bus fleet is now able to run cleaner than brand-new buses. "Part of the reason behind the decision to use soy-based biodiesel is that we're always concerned about emissions, and this helps us help the environment," said Roger Hageman, FST&P maintenance manager. "We are very conscientious about supporting Iowa's farmers, and using B20 will benefit them while decreasing our dependence on foreign oil."[12]

Saint Louis, Missouri, has been the location for another long-running biodiesel transit bus program. For more than ten years, the Bi-State Development Agency, a mass transit provider for the Saint Louis area, conducted extensive testing of biodiesel for the U.S. Department of Energy and the National Renewable Energy Lab. The results of the tests were favorable and demonstrated a significant reduction in vehicle emissions without an impact on fuel economy or performance. Use of the B20 fuel posed no operational problems in the transit buses that participated in the tests. Bi-State (now called Metro) noticed that in addition to reducing vehicle emissions and the release of particulate matter, B20 had such good lubricity that it increased injector life and decreased the need for vehicle maintenance, a finding that has generally been confirmed by other fleet tests. Passengers riding the B20-fueled buses appreciated the absence of the acrid smell and black exhaust smoke normally associated with diesel buses. Metro plans to incorporate biodiesel into its entire fleet of diesel buses.[13]

Diesel-powered transit buses are going to be around for many years to come. In order to meet the new stringent EPA low-sulfur rules that will go into effect in 2006, many transit authorities are switching to ultra-low-sulfur fuel and are installing diesel particulate filters. Switching to biodiesel is an extremely cost-effective method of meeting the new EPA rules while potentially reducing maintenance costs and possibly increasing mileage.

Biodiesel and Uncle Sam

From the U.S. Marine Corps Base in Camp Lejeune, North Carolina, to Everett Naval Station in the Puget Sound area of Washington State, military installations across the nation are using biodiesel—and a lot of it—in their diesel-powered vehicles. The U.S. military, in fact, is the single largest user of biodiesel in the country, consuming about 5 million gallons annually. The U.S. Army, Navy, Air Force, and Marines all use B20 in their nontactical vehicles. Of the four branches of the military, the Marine Corps uses B20 in the largest number of locations. "We use biodiesel to help us meet our federal alternative-fuel requirements and to reduce our petroleum fuel consumption to meet the Executive Order directing the government to do so, and on a third level it is just the right thing to do," said Tim Campbell, Headquarters Marine Corps GME program manager. "We've had no reported maintenance issues. I asked the bases to contact me with their experiences, negative or positive, with biodiesel. I received only positive feedback." Most of the military installations get their biodiesel through the Defense Energy Support Center (DESC), which coordinates the federal government's fuel purchases. DESC is the largest single purchaser of biodiesel in the country and for the 2003–2004 period had requirements for 5.2 million gallons of B20 for both military and civilian locations across the country.[14]

The U.S. Marine Corps Base at Camp Lejeune, North Carolina, has used biodiesel for about two years in approximately three hundred to four hundred pieces of equipment, including buses, caterpillar tractors, and bulldozers, and consumes approximately 148,000 gallons of B20 a year. Scott Air Force Base in Illinois, located about 30 miles east of Saint Louis, Missouri, serves as headquarters for twelve Air Mobility Command bases throughout the nation. Scott AFB has used B20 since April 2001 and consumes about 75,000 gallons every year. Located in the Puget Sound area, the Everett Naval Station in Everett, Washington, has used about 50,000 gallons of B20 annually since 2001. The change to biodiesel was virtually seamless, according to transportation director Gary Passmore. "Older equipment took a filter change, but newer equipment needed nothing," he said. "It went so smoothly that no one really noticed."[15]

The Navy probably deserves the credit for one of the most interesting recent biodiesel developments. On October 30, 2003, there was a ribbon-cutting ceremony at Navy Base Ventura County in Port Hueneme, California, that could have big implications for both the biodiesel industry and the military. The ceremony, attended by about a hundred people, marked the unveiling of the Navy's first mini-refinery for converting used restaurant cooking oil into biodiesel. The refinery, developed by Biodiesel Industries of Santa Barbara, California, in cooperation with the Naval Facilities Engineering Service Center (NFESC) at the base, is small enough to fit on the back of a pickup truck and can produce the fuel in 200-gallon batches. "This is the culmination of four years of working with the U.S. Navy," says Russell Teal, CEO of Biodiesel Industries. "Our research and development of the Modular Production Unit has been completed and implemented in our civilian plants in Las Vegas and Australia. Now, with the cooperation of NFESC, we hope to continue making improvements so that it can soon be deployed at military installations around the world."[16]

Since the official ceremony, the base has been collecting used cooking oil and transforming it into biodiesel fuel for use in its vehicles at the naval facility. The base intends to use about 20,000 gallons of biodiesel in its own vehicles annually and will also be producing about 20,000 gallons each for the nearby Channel Islands National Park and for Ventura County. The mini-refinery not only offers the naval facility a convenient way of disposing of a solid waste product but also provides a measure of energy security, according to Kurt Buehler, chemical engineer at NFESC. "If petroleum gets cut off, we can keep the base running on biodiesel," he notes. "So, in addition to reducing dependence on foreign oil, producing our own biodiesel could provide a tactical advantage in case of crisis."[17] This same idea is beginning to occur to planners and commanders throughout the U.S. military, and this creates a huge potential market for biodiesel and biodiesel technology providers. The concept of an "all-American" fuel in the national defense sector is a compelling argument.

Biodiesel in the Park

The highly publicized 1994 "Truck in the Park" project conducted in Yellowstone National Park (the first national park in the nation to test biodiesel) has led to additional biodiesel use in the park. The project has now been expanded to include tour buses, garbage trucks, and heavy equipment as well as boilers. Beginning in the spring of 2002, all of the park's three hundred diesel-powered vehicles began running on B20. In October of the same year, the first public biodiesel pump in Montana was opened at the Econo-Mart in West Yellowstone. In addition, it is intended that biodiesel will be made available to the general public at all gas stations within the park itself. "We're stewards protecting this national treasure, and using biodiesel is one way we can best do that," said

Jim Evanoff, management assistant at Yellowstone. "I have talked with hundreds of visitors about biodiesel use at the park, and the majority of our visitors are really interested in renewable fuels. This is an educational program as much as it is an environmental one."[18]

But Yellowstone is not the only national park to adopt biodiesel. Mammoth Cave National Park in south-central Kentucky is the first national park to have virtually all of its vehicles powered by alternative fuels. All of the park's transit and support vehicles run on either biodiesel or ethanol. The park's tractors, backhoes, graders, and even riding lawn mowers are running on B20. Two ferries on the Green River, which carry about three hundred cars and light trucks each day during the summer, run on biodiesel as well. The park's tour buses, however, run on propane. The 10,000 to 12,000 gallons of biodiesel consumed by the park annually are supplied by Griffin Industries of Cold Spring, Kentucky.[19] The biodiesel projects at Yellowstone and Mammoth Cave have been great successes with park employees and visitors alike, and the National Park Service has since introduced biodiesel to a total of twenty parks across the country—including Harpers Ferry National Historic Park, Everglades National Park, Glacier National Park, Yosemite National Park, Hawaii Volcanoes National Park, and many more—through the Green Energy Parks Program.

One of these parks, the Channel Islands National Park, located off the coast of Southern California, has been able to demonstrate the many benefits of using biodiesel in a marine environment. The park consists of five islands and the surrounding mile of ocean, totaling 249,489 acres. The islands' isolation has protected them from development but presents park management with a real challenge in maintaining energy services. But the greatest challenge of all was marine transportation. The park's boat fleet used more than 70,000 gallons of diesel fuel annually. In August 2000 the park implemented its biodiesel program and now uses B100 in the vessels *Pacific Ranger* and *Sea Ranger II*, as well as in diesel equipment, including stationary power generators and forklifts, on the islands. The use of biodiesel and other renewable resources makes the islands petroleum free. "We are an environmental organization, and we should be willing to be in the forefront in demonstrating things that have a positive environmental impact," said Kent Bullard, maintenance supervisor at the park. "It has been seamless. We haven't had any performance issues; the biodiesel is performing just as well as diesel."[20]

In December 2003 Yosemite became the first national park in the country to produce its own biodiesel on-site with the arrival of a small processing unit developed by Biodiesel Industries of Santa Barbara, California. Delaware North Parks and Resorts, the concessionaire that provides guest services in the park, now uses the processor to convert used restaurant cooking oil into biodiesel for Park Service vehicles.[21] Other parks with restaurant facilities are expected to follow this example in the future.

Municipalities Go Green

For quite a few years many cities and towns across the country—Keene, New Hampshire; Takoma Park, Maryland; Columbia, Missouri; Breckenridge, Colorado; and Missoula, Montana, to name just a few—have been switching some of their vehicles to run on biodiesel blends. But the city of Berkeley, California, has taken the switch to biodiesel to its highest level. In 2001, thanks to strong encouragement from the Berkeley Ecology Center, a local community and environmental organization, the city became the first to run its entire fleet of recycling trucks on B20. The success of the switch impressed city officials, and they decided to expand the initiative to virtually all of the city's diesel-powered vehicles, including fire trucks, school buses, and public works vehicles. In 2002 the Ecology Center upped the ante by switching to B100 use in its vehicles and encouraging the city to do the same. On June 24, 2003, the city held a ceremony officially announcing its transition to B100 in its *entire* diesel fleet, making Berkeley the first city of its size in the nation to do so. "The city of Berkeley has a long history of innovation and as a leader in public policy," said City Manager Weldon Rucker. "The use of biodiesel fuel is yet another example." Berkeley now uses B100 in more than 180 of its diesel vehicles, representing 90 percent of its fleet of two hundred diesel vehicles. The remaining 10 percent are fire department vehicles that will also be switched to B100 after deliveries of the fuel can be worked out for the city's more remote fire stations.[22]

Biodiesel Goes to College

While there has been a good deal of biodiesel research at a number of colleges and universities in the United States, institutions of higher learning in general have been slow to adopt the use of biodiesel on their campuses. In February 2004, however, Harvard University in Cambridge, Massachusetts, announced that it had begun to use B20 in all of its diesel vehicles and equipment, including shuttle buses, mail trucks, and solid waste and recycling trucks. Although a number of alternative fuels were studied, biodiesel was finally selected because it provided the greatest health and environmental benefits in the most cost-effective way, according to David Harris Jr., general manager of transportation services at Harvard. But there were other reasons for the switch as well. "Harvard is not a stand-alone campus," Harris said. "Our shuttle buses drive down the streets of Cambridge, past houses and other schools. We feel a responsibility to be a good neighbor and be as environmentally friendly as possible. Biodiesel helps us accomplish that using the vehicles we already have."[23] World Energy of Massachusetts supplied the biodiesel.

In April 2004 Purdue University in Indiana announced the introduction of B2 in as many as eighty of its vehicles, including nine university buses and dozens of diesel trucks. "Our decision to use biodiesel represents a balance between supporting the ag community and also keeping our diesel fuel costs reasonable," said Mike Funk, Purdue's director of transportation, about the relatively modest amount of biodiesel involved in the switch.[24] Yet Purdue's decision is definitely a step in the right direction, one that other institutions could easily follow without causing any major financial problems. Other institutions of higher learning using biodiesel include the University of Colorado; New Hampshire's Keene State College; North Carolina State University; the University of South Carolina; Clemson University; Indiana University; the University of Michigan; Northwest Missouri State University; and, of course, the University of Idaho, where U.S. biodiesel research began.

Biodiesel on the Slopes

In January 2001 Aspen Skiing Company released its *Sustainability Report*, the first formal sustainability report in the industry, which took a candid look at the company's environmental impact and explained what it was doing about it. Every year the company uses about 260,000 gallons of diesel fuel for activities like trail grooming by snowcats. The emissions from these tracked vehicles create a good deal of air pollution, and Aspen decided immediately to begin the switch to B20. After initial tests proved to be a success, Aspen switched its entire fleet of snowcats to biodiesel in the winter of 2003. Eventually Aspen hopes to make biodiesel from used cooking oil from its restaurants.

In February 2004 New Hampshire's Cranmore Mountain Resort announced that it was switching to biodiesel to power all of its snow-grooming machines. The Cranmore Mountain biodiesel project is a collaboration between Cranmore, the New Hampshire Department of Environmental Services, and the Granite State Clean Cities Coalition. The 5,000 gallons of fuel used monthly in the new initiative were provided by World Energy of Massachusetts. Cranmore was the first eastern winter resort to adopt the use of biodiesel. "Cranmore is passionate about taking measures to help the environment," said Jim Mersereau, mountain manager for Cranmore Mountain Resort. "We are proud to be the first resort in the East to use this alternative fuel."[25] Last but not least, the Breckenridge Ski Resort in Colorado has also been a leader in adopting the use of biodiesel, and has even hosted a workshop for fleet managers who were interested in learning more about the fuel. So much for concerns about the viability of biodiesel in severe winter-weather conditions.

Biodiesel Dragsters

The speed with which biodiesel use has been growing in the United States has been exceeded only by the speed achieved by a biodiesel-fueled dragster, which set the world renewable-fuel speed record using B100. On September 14, 2002, driver Mark Smith pushed "Wild Thang" to 211 miles per hour on the 660-foot racetrack at the Ozark International Raceway in Rogersville, Missouri. The dragster used biodiesel produced by West Central Soy of Ralston, Iowa, to achieve the Guinness World Record. "The biodiesel definitely impressed me," said Russel Gehrke of Seymour, Missouri, who helped prepare the car for its run. "It ran just as fast as conventional, but much cleaner. The crowd really liked it, too. It smells a lot better than diesel. When you're burning a gallon of fuel per second like we are, you want something that is environmentally friendly. I would definitely use biodiesel again." The National Biodiesel Board, of course, was excited about the new record. "Achieving a speed of 211 miles per hour with biodiesel just underscores that this is a high-performance fuel," said Joe Jobe, NBB executive director.[26]

In September 2003 a 2001 Volkswagen Jetta TDI running on B100 captured an impressive array of performance awards at the 2003 Michelin Challenge Bibendum held at Infineon Raceway in Sonoma, California. Michelin bills the Challenge Bibendum as the largest environmental vehicle event in the world. It is considered a performance event rather than a competitive event and is intended to display advancements in vehicle technologies. Consequently, entrants are rated only with A, B, C, and D letters. The Jetta, entered by American Biofuels, captured more top ratings than any other production class vehicle, earning an A rating in six categories, including energy efficiency, carbon dioxide, and range. It achieved more than 60 miles per gallon while clocking some of the fastest lap times in the fuel efficiency event. "Overall, the progress towards sustainable mobility by all of the participating technologies and energy sources is very impressive," said Patrick Oliva, director of Challenge Bibendum. "Each year, the variety of technologies and creative innovations displayed offer proof that sustainable mobility is within our grasp."[27]

PART FOUR

Biodiesel *in the* Future

12

Looking Ahead

In June 2004 the cost of gasoline reached a record-high price of $2.06 per gallon on average across the United States, and in October U.S. light crude hit $55.67 a barrel, the highest level in twenty-one years of recorded oil prices. Just a year earlier light crude had been selling for $29.21 a barrel. These dramatic price increases were due to a wide range of domestic and international issues, but especially to the deteriorating military and political situation in Iraq, a number of terrorist attacks on Western oil workers in Saudi Arabia, and increasing instability in the Middle East in general. Oil traders don't like instability of any kind, and the ongoing chaos in the Middle East was giving them fits.

In the United States, the steady rise in gas prices at the pump has sparked increased interest in hybrid-electric as well as diesel-powered vehicles. As waiting lists for Toyota's popular 2004 Prius grew longer, the wait for actual delivery extended to six months or more for many frustrated buyers, while sales of diesel-powered vehicles also began to accelerate in some parts of the country. And many people who bought those diesel cars or light trucks were planning to use biodiesel as fuel, according to some dealers.

What does all this portend for the biodiesel industry? While it's difficult to predict exactly where petroleum prices will be in the future, it's a fairly safe assumption that if demand remains strong, especially in countries such as China (which most industry analysts had underestimated), and reserves and production capacity remain limited (which seems likely), there will be a general upward trend in prices in the years to come. While some oil industry analysts insist that current high prices are just a temporary situation, others have stated recently that oil prices possibly have reached a new plateau and may not come down much at all, and that further price increases are inevitable. If this turns out to be true, the tipping point for petroleum may not be as far off as many had predicted.

But bad news in the petroleum sector is generally good news in the biodiesel sector. Assuming that petroleum prices remain high and that the recently passed tax incentive for biodiesel boosts the industry as expected, the future of biodiesel in the United States looks promising. In Europe, with the addition of the new accession countries on May 1, 2004, huge new areas of agricultural lands that could be planted with biodiesel feedstocks became available to the new EU-25—to say nothing of the addition of millions of potential new biodiesel customers. That, coupled with a solid, well-established, technically advanced industry makes for a very favorable situation for European biodiesel producers and consumers alike. And elsewhere around the world, as more and more countries discover the many advantages of biodiesel, from increased energy security to more local jobs and environmental benefits, the prospects for increased growth in the biodiesel sector are equally encouraging.

Key Issues

Nevertheless, a number of issues need to be addressed if the industry is going to achieve its full potential. Raffaello Garofalo, the secretary-general of the European Biodiesel Board, has some thoughts on these issues that he shared at the 2004 Biodiesel Conference & Expo in Palm Springs, California. Although his presentation was about the EU biodiesel industry, his comments could easily apply to much of the industry around the world. "I presented three key concepts on the future development of biodiesel," he says. "First is quality. Quality is really crucial because consumers purchase biodiesel not only because it is good for the environment, but because it performs well in their engines. Quality is also the way to gain acceptance by the auto industry as well as by our customers."[1]

The second concept Garofalo presented is what he refers to as "progressive development," which relates to a steady, careful, long-term strategy for the industry. "Biodiesel should not be seen as simply an adventure by people who produce biodiesel in their basement," he says. "It requires huge investments, as well as a long-term vision with big companies creating a real industry. The petroleum industry requires us to provide them with large quantities of biodiesel at the times and places where they need it. And that will take steady, progressive development instead of short-term considerations such as building a biodiesel plant in order to get subsidies and then ignoring quality; this is the worst enemy of biodiesel."

Finally Garofalo described what he calls the "complementary approach" for the industry. "This means that biodiesel cannot be developed in opposition to other industries," he says. "Biodiesel needs to be combined with the contributions of all the other stakeholders, including the petroleum industry; we have to work together with all of them. The same applies to original equipment manufacturers, and of course we have to work with public authorities in order to get the taxation support that we need. I am convinced that the success of the industry requires the contribution and cooperation of everyone, from public authorities, car manufacturers, the petroleum industry, the vegetable oil industry, the consumer, and even the other biofuels industries. Sometimes in Europe you hear people saying that bioethanol is better than biodiesel or the opposite. This is not helpful for us or for them. If we cooperate with all of these stakeholders, it will give biodiesel a better chance to develop, not only in Europe but worldwide."[2]

Many people in the industry also admit that there is a need for greater cooperation within the industry itself. The 2004 Biodiesel Expo provided a venue to begin to address that issue, and many others as well. The Expo was a unique opportunity for about five hundred participants to attend presentations, take part in panel discussions, and have one-on-one conversations with representatives of virtually every segment of the industry. The event highlighted for participants a number of key insights about the present state of the industry.

- The biodiesel industry has been growing at a remarkable rate, especially in the past few years.
- The industry is diversifying and developing a broad range of new products, many of which were on display at the gathering.
- More research and development information is available than ever before, providing the industry with an invaluable array of resources.
- The National Biodiesel Board is becoming receptive to new ideas that should allow the organization to evolve along with the industry.
- More work still needs to be devoted to marketing and promotion to explain biodiesel to the vast majority of the general public that still does not really understand what biodiesel is or what it can do.

Almost everyone who attended agreed that the gathering was an exciting and valuable event. But for some the Expo bordered on an emotional experience, particularly for those individuals who have been involved with the industry from the early years. Bill Ayres, formerly with Midwest Biofuels and Ag Environmental Products, attended the gathering along with Kenlon Johannes, CEO of the Kansas Soybean Commission & Association. (Johannes, a tireless supporter of biodiesel for many years, was the first executive director of the National SoyDiesel Development Board.) "Kenlon Johannes and I were sitting in the back of the room listening to Daryl Hannah and all these top people from the USDA and DOE talking about biodiesel," Ayres relates, "and it was kind of like watching your kid graduate from college; the industry really has matured and has so much going for it now. It's really exciting to see the growth in the industry, and I think it's going to do quite well in the future."[3]

Quality

The fact that quality was at the top of Raffaello Garofalo's list of key issues for the European biodiesel industry is no accident. Virtually everyone in the industry around the world acknowledges the central role that quality needs to play in the future of biodiesel. But it's also a fact that quality—or lack thereof—has occasionally been an issue for many producers, both large and small. Over the years, there have been repeated reports of off-spec biodiesel being made by producers in many different countries. In some instances there may have been considerable justification for the claims, while in other cases the rumors appear to have been totally false. Regardless of the particular circumstances surrounding these incidents, bad biodiesel is bad for the industry. "The standard myth from the industry's point of view is that big producers make better fuel than small producers," says Tom Leue, president of Homestead Inc. in Williamsburg, Massachusetts. "But the reality is that they don't always do that. Nevertheless, small producers do need to 'get the religion' of quality control; we simply have to make sure that we produce quality biodiesel. There is no point in producing bad fuel, it doesn't help anybody."[4]

"Quality is definitely job one," says the National Biodiesel Board's executive director Joe Jobe. In response to ongoing quality concerns, the NBB is instituting in the United States a new quality-assurance program called BQ-9000, Quality Management System Requirements for the Biodiesel Industry. Developed and implemented by the National Biodiesel Accreditation Commission, an autonomous nine-member group appointed by the NBB, the voluntary program is designed to ensure more consistent quality for biodiesel from producer to end user. There are two levels of certification: accredited producer and certified marketer. "You can apply for both if you are a producer as well as a marketer," Jobe explains. "If you are a certified marketer and buy from an accredited producer, then you can rely on some of the assurances of the accredited producer. But if you are a certified marketer and you buy from a nonaccredited producer, then the burden of testing and storage and so on is your responsibility. It's a pretty sophisticated program that is based on typical ISO 9000–type programs that are prevalent in almost any manufacturing industry. We already have a specification that assures quality biodiesel production. But BQ-9000 is intended to ensure quality from the plant gate to the fuel tank, because once biodiesel leaves the plant there are all kinds of things that can happen to it before it gets to the end user that can cause problems."[5] Germany's Association for Quality Management of Biodiesel (AGQM) and similar organizations in other countries around the world reflect the widespread recognition of the importance of quality control to the future success of the industry.

Small Regional Markets

Because biodiesel is so easy to make on a small scale, there are numerous possible models that could be followed for local production by small organizations. "I think that it would be relatively easy for almost every county in the nation to take a lot of their used cooking oil and even some of their virgin oils and use them in the production of biodiesel for their own fleets of vehicles," says Daryl Reece, vice president of engineering at Pacific Biodiesel. "Sure, it's going to cost a few pennies a gallon more to do that, but it's a good alternative to petroleum-based fuels. You don't have the transportation costs because everything's right there. And you can produce it with one or two people, and then it's utilized locally. Some of these municipalities could even use biodiesel to generate their own power. There are very good tax incentives for green energy power generation right now. There's a lot of things that could be looked at, and they're all very positive."[6]

Following a similar line of thinking, planners at the University of Missouri in Columbia, Missouri, have suggested the creation of what they call "community-size" biodiesel hubs. This strategy involves the production and use of biodiesel within regional agricultural/industrial centers or hubs. The production level for this model is in the 500,000-gallon-per-year range. This strategy, however, would require a good deal of planning and cooperation among all the participants in order to succeed.[7] In a variation on this theme, some independent producers are beginning to focus on niche or small regional markets that emphasize ecologically sound practices, such as avoiding the use of genetically modified crops and pesticides to grow bioenergy crops or using waste vegetable oil as a feedstock. By focusing on biodiesel as an organic and/or locally produced product that also helps stimulate the creation of local jobs, these producers are able to offer their environmentally savvy customers an attractive alternative to the large agribusiness model.

This general model is similar to what Co-op Plus and Homestead Inc. of western Massachusetts, who plan to use waste cooking oil as a feedstock (see chapter 9), have in mind. "I think it is inevitable that we start producing locally based fuel and that we utilize what are currently badly managed resources," says Tom Leue, Homestead's president. "I also think it is inevitable that biodiesel will take its rightful place on a solid footing alongside our dwindling petroleum resources. I don't think that biodiesel is the salvation of the world, simply because there isn't enough. But biodiesel is for those individuals and communities that can come to grips with their energy-use problems and who strive toward energy independence. They may not be able to get there completely, but at least they can move toward it. I don't have any illusions that we are going to solve all our problems with biodiesel, but I still think we've got to make the effort anyway. And I think that we can do it better than has been done to date by keeping it in the hands of local people rather than big industry."[8] Similar initiatives are being planned in other parts of the country by various groups, to say nothing of locally focused initiatives like the nanohana projects in Japan or the community-based initiatives in India and South Africa (see chapter 7).

Growing a New Energy Economy

A few optimistic observers feel that the United States has sufficient resources from fallow farmland, waste cooking oil, and all other sources to produce enough fuel to meet as much as 25 percent (13.8 billion gallons) of the nation's diesel needs. Most estimates, however, are more conservative and put the figure somewhere between 14 and 20 percent. Even more conservative observers say the figure is closer to 6 percent (3.3 billion gallons), given current agricultural practices. Pioneer biodiesel researcher Dr. Charles Peterson, now an emeritus professor of biological and agricultural engineering at the University of Idaho, isn't convinced that even the latter projection is realistic. "In terms of total production, I think 3 to 5 percent of our current diesel use would be very ambitious," he says. "I did a study on the development of the biodiesel industry and projected that if we could ever get up to 1 billion gallons, that would be remarkable. I think that 500 million gallons would be a huge market in the U.S. and that somewhere around 150 to 200 million gallons is where we will really test whether biodiesel can continue to grow."[9] The debate continues.

In a 2002 German study on international biodiesel production potential, it was estimated that the upper limit of biodiesel replacement of petrodiesel worldwide was around 10 percent.[10] In some countries that figure might be lower, while in others it might be higher, depending on available feedstock and other factors, according to Werner Körbitz, chairman of the Austrian Biofuels Institute. "For example, Malaysia, which is producing a lot of palm oil, could reach a higher market share than, say, Egypt, a country that does not produce much oilseed," he says. "In Europe, including eastern Europe, and North America, I think reaching 10 percent would be a challenge. Brazil and Argentina, on the other hand, might have an easier time since they grow a lot of soybeans. But then it all depends on whether you make more money selling the oil for food purposes or if you have surplus to sell to make biodiesel."[11] Regardless of which figures are correct, it's clear that biodiesel is never going to be a total replacement for all of the current petrodiesel market.

New Feedstocks

Trying to deal with that reality is a challenge. In the United States, simply requiring more fuel-efficient vehicles would almost certainly result in substantial reductions in petroleum use, but this is mainly a political, rather than a technical, problem. It's clear that U.S. oil dependence has become so pervasive that it cannot be resolved without a combined strategy of various alternative energy sources in addition to conservation. It's also increasingly clear that if biodiesel is going to play a significant role in this scenario, the U.S. industry needs to lower the cost at the pump as much as possible. And since about 70 percent of the cost of production is represented by feedstock, the cost of feedstock is the key factor that needs the most attention. If the right feedstocks were being grown on the available land resources, biodiesel production could be significantly increased while its price could be simultaneously reduced. "For the consumers of the 55 billion gallons of diesel sold in this country every year the main issues are price, price, and price," says Jerrel Branson of Best BioFuels in Texas. "If the price was right, biodiesel would be at a pump near you tomorrow."[12] Jon Van Gerpen at Iowa State University agrees. "Quite simply, biodiesel has been too expensive," he says. "If it had been a low-cost fuel, of course it would have happened immediately; people would have bought all that the industry could make. But it has always been expensive, and so it's been an uphill battle."[13]

Since cost is such a key factor, people both inside and outside of the biodiesel industry are beginning to question the wisdom of continuing to focus so much attention on a feedcrop—the soybean—that is not very high on the list of oil-producing seeds. "Given the very high price and low oil yield of soybeans, it's just not economic to even think about making biodiesel out of soybeans," Jerrel Branson maintains. "This is not the raw material that will make a competitive product."[14] Dennis Griffin, chairman of Griffin Industries in Kentucky, agrees. "Using an edible oil to produce fuel has never made a lot of sense to me," he notes. "That's why we wanted a multifeedstock type of production plant, so that as new, more cost-effective feedstocks are found we can adapt to them—and they certainly are not going to be in the edible oil class."[15]

While most soybean farmers probably would take exception to these statements, nevertheless a lot of them recognize that, at the very least, the industry needs to further diversify its selection of feedstocks. And these farmers are not necessarily opposed to this initiative, according to NBB's Joe Jobe. "Soybean farmers have had a very broad vision of an inclusive industry," he says. "They are often criticized for just the opposite, but I've been around this for seven years, and they have built an industry that is ultimately based on multifeedstock markets." What's more, Jobe believes that as the biodiesel industry continues to expand, alternative feedstock crops such as canola and mustard will almost certainly play an increasingly significant role in the United States. "I think we would have an agricultural response to that growth, so that more oilseed varieties would be added to both cultivated and fallow acres," he predicts. "There is already some of this sort of response going on in Colorado, Idaho, Hawaii, and other places in order to create higher-yielding oil commodities. There's a tremendous opportunity to switch wheat acres and other crops to sunflower and rapeseed and so on; there's a lot of room for that. And this is part of the anticipated agricultural response that would have a pretty substantial impact on our feedstock capabilities."[16]

Additional evidence that an agricultural response is already underway can be found in the Pacific Northwest, where rapeseed for biodiesel has been suggested as a good rotation crop for wheat farmers in Oregon and Washington. In Colorado, mustard seed is actively being tested as a rotation crop by farmers in cooperation with Blue Sun Biodiesel. But the company's feedstock research is not limited to mustard seed, according to Jeff Probst, president and CEO of Blue Sun. "There are other technologies that need to be investigated and worked on, like algae," he says. "We can grow algae on a very cost-effective basis, but there is still a lot of development that needs to go on. People talk about these feedstocks now, but I think the private sector can actually make them happen. That's the really exciting part of this business; we don't have to rely on soybean oil. We want to grow our own industrial oilseed that doesn't have an impact on the food supply, and I think there is a lot more that can be done with that."[17]

The increasing role that the private sector is beginning to play in biodiesel research and development is a significant shift for the industry, which has relied on a good deal of government-supported research for many years. Admittedly, some of this new research is still partly funded by government grants, but industry's growing interest in biodiesel is viewed by most observers as a positive development. "One of the exciting things that's happening right now is that industry is beginning to see that there is a real benefit to using biodiesel," says Professor Leon Schumacher, a long-time biodiesel researcher at the University of Missouri in Columbia. "So they are getting involved in some of the research now, which is a good thing. They can really focus some effort on problems that perhaps haven't been solved previously and give them the extra push they need."[18]

A Few Concerns

Despite all the enthusiasm in the industry for various new feedstock crops, some observers—especially some in the environmental community who otherwise are in favor of biodiesel—are troubled by the fact that corporate agribusiness is so heavily involved in supporting increased biodiesel use, and they are concerned that genetically modified crops will be used. Dr. Rico Cruz, of the Environmental Science and Technology Program of the Confederated Tribes of the Umatilla Indian Reservation in Washington State, has worked hard to establish small-scale biodiesel initiatives around the world and shares some of those misgivings. "I am concerned about the development of a monoculture," says Dr. Cruz. "What if there is a disease that wipes out the whole crop? My concern is about relying on one or two genetically engineered crops while having some other crops that have been here for thousands of years disappear because farmers have signed a contract with one of these big companies."[19]

But some nagging concerns extend beyond the realm of genetic engineering. If biodiesel production and consumption continue to grow rapidly, additional questions will almost certainly need to be asked about the ecological implications of using food resources—all these soybeans and other edible-oil crops—for fuel rather than for food. Similar questions could be raised about the use of land and agricultural resources for the cultivation of inedible oil crops, unless they are grown on fallow or marginal lands using ecologically sound agricultural practices. In some parts of the world, biodiesel strategies are based primarily on the use of inedible crops grown on set-aside or marginal lands, eliminating or at least mitigating some of the concerns about using resources needed for food production, agribusiness involvement, and potential problems with land degradation. The food versus fuel debate has been ongoing for many years, and although it raises important concerns, in general they may have been exaggerated. The debate relates to an extremely complex series of issues that extend well beyond the scope of this book (for an excellent in-depth discussion of these issues go to the Journey to Forever Web site at journeytoforever.org/biofuel_food.html).

Another major challenge for the industry is trying to figure out ways of mitigating the gut-wrenching price swings in feedstock commodities that tend to cause so much instability in the biodiesel market. "In both Europe and the United States, we have had these wild swings between profitability and loss," says Gene Gebolys, president of World Energy in Massachusetts. "This basically means you have a very hard time growing the capacity of the industry properly, because you keep on losing companies during long periods of loss, and then you have wild boom periods where everybody wants to build. This has already happened in Europe, and it's not a good way to grow capacity or grow a business. Right now, the way it is, it's really unstable."[20] Developing a broader array of feedstocks is one way of smoothing out at least some of the larger bumps in the commodity markets.

The Next Level

Although many observers feel that the U.S. biodiesel industry is about ten years (some say twenty) behind the European biodiesel industry, there is no question that the industry in North America is rapidly catching up and represents the fastest-growing segment in the global arena. "The long view that has been taken by the farmers who supported the industry for so many years in this country has proven to be successful," says Joe Jobe, "because around 1999 we finally began to make the transition from the research and development phase to the commercialization phase. Since then, we've seen double or triple growth every year for the past four years in a row. The momentum and enthusiasm for biodiesel is growing, and the next challenge will be to get us from where we are now to the next level."[21] That next level, according to Jobe, includes broader public acceptance as well as the integration of biodiesel into future energy strategies at the governmental level and with the fuel and engine manufacturers. "Our overall broad vision for the industry involves varied markets and continued growth. The B20 market is currently our single largest market," Jobe continues, "but we also see a long-term growth in low-blend markets, because B2 or B5 is something that we can do as a nation to displace up to 5 percent of our diesel fuel. With the help of the tax credit, we can achieve those goals and easily displace 5 percent of petrodiesel fuel needs by growing it right here in the U.S. I'm very proud of our accomplishments, and we need to keep working hard on achieving even more."[22]

The rapid growth in the industry around the world—which in recent years has been breathtaking—has left little time to make sure that the growth is orderly and well managed. Some observers are concerned that this can lead to problems. "We have to be careful that we don't go too fast," cautions Professor Leon Schumacher. "Any industry that is growing this quickly needs to try to control the process and make sure that the things that need to be in place are there. For example, do we have adequate infrastructure and trucking? Can we get the product? That may sound obvious, but here in Columbia there have been times when we just couldn't get the product. No amount of advertising is going to move the industry ahead if its product is not available. Those are the sort of things that have to be factored in as we move ahead."[23]

Although many in the industry welcome any support they can get, the heavy reliance of the biodiesel industry on government subsidies, tax breaks, and research support is a matter of some concern. And those concerns are only going to increase as the industry moves on to the next level of expansion around the world. "Not only in the U.S., but also in Europe, we are dependent on public support, and maybe too much on that support," says Raffaello Garofalo. "That will be a big challenge for the industry in the next ten to twenty years."[24] Jeff Probst agrees, saying, "I think the greatest challenge is to get out of the public sector and into the private sector. Everybody looks at the private sector as the bad guys who are concerned only about profits, but that's not true. The private sector is just as interested in cleaning up the air and getting ourselves weaned from imported foreign oil as any government official. What we really need is better public and private cooperation, but right now the private sector is not being regarded as a key player in the progress of alternative fuels and biodiesel."[25]

The Best Alternative

Virtually everyone in the industry around the world readily concedes that there isn't enough feedstock to allow biodiesel to completely replace petrodiesel. Some actually see this as a good thing, since replacing one total energy dependency—petroleum—with another—biodiesel—would hardly be an improvement. The primary strategy for the industry is to identify and fill key niche markets where biodiesel's use will do the most good, while reducing reliance on at least some petrodiesel. In addition, most in the industry would also admit that biodiesel isn't perfect; slight increases in NOx emissions as well as some cold-weather issues are probably the fuel's main disadvantages. But a number of strategies have been identified in recent years to substantially mitigate these problems.

"While biodiesel isn't the total solution, I think it probably is the best alternative fuel we have at the moment," says World Energy's Gene Gebolys. "If there is a broad recognition that making positive steps in the right direction with renewable fuels is something that is important, then biodiesel is going to succeed." Along the way, Gebolys predicts that there will be some rough spots, with periods of overbuilding or underbuilding of production capacity and instances of well-intentioned government policy that will have unintended consequences and that may sometimes turn out to be counterproductive. "So it's not always going to be pretty, but on balance, I think we are going to see generally steady growth of the industry," he says. "The biggest obstacle is inertia. We have to constantly try to move people away from what they did yesterday. I wouldn't say that there are too many obstacles in this industry that are unique or much different from those of any other brand-new industry."[26]

Bob King of Pacific Biodiesel is optimistic about the industry's prospects, too, and says that many people are finally connecting the dots about petroleum dependency and foreign and domestic government policy. "I have one customer who has a bumper sticker on his vehicle that says 'Biodiesel: No War Required,'" he reports. But King is enthusiastic about biodiesel for many other reasons as well. "I understand economics, and I look at all the different alternative energy concepts that are out there—hydrogen fuel cells, ethanol, solar, wind—and I am just totally excited about biodiesel," he declares. "All those other fuels will be part of that future too, in their own niche, because we need all of them. But biodiesel just has so much going for it. It's easily produced and is used so efficiently in the diesel engine with technology that we already have. We hope to be long-term players in the future, but whatever happens I'm totally convinced that biodiesel is going to be a major portion of our energy future, so we're going to do what we can to make that happen. We're not hopelessly locked into a petroleum future if we don't want to be."[27] That's a good thing, since petroleum's future is looking increasingly unstable and unpredictable.

Gary Haer of West Central Cooperative in Iowa is also enthusiastic about the industry's future prospects. "I think that the biodiesel industry is poised for significant growth," he declares. "The product has exceptional lubrication and performance properties, is easy to use, and can be delivered through the existing infrastructure for petroleum fuels. And it's a reliable, renewable alternative fuel that comes from domestically produced resources. It just makes a lot of sense. We're able to keep our energy dollars in our own country instead of exporting them overseas to buy crude oil from foreign sources."[28]

In some countries, biodiesel offers additional advantages beyond good performance in diesel engines and energy security. "In our country, it's all about employment and empowerment for the low-income groups," says Darryl Melrose, owner of Biodiesel SA in South Africa. "Each country has its own initiatives and pros and cons regarding this fuel. In our case, we don't have a major air pollution problem, but creating jobs is an important issue here, and jatropha will play a big role. I don't see biodiesel being sold much cheaper here than regular diesel, and it probably won't have a major impact on our usage, given the amount of land we have available and our consumption. But it certainly will help with the bigger picture regarding employment."[29] Other countries, such as India, with large numbers of relatively impoverished rural populations see similar potential advantages with biodiesel.

A Bridge to the Future

Many observers view biodiesel as a "bridge technology" that will help get the world from where it is today—almost totally dependent on fossil fuels—to where it needs to be in the future—relying on a broad range of renewable energy strategies, conservation measures, and zero-emissions vehicles. Biodiesel is an immediate solution that fits current infrastructures, including those of highways, distribution networks, and vehicle fleets. It runs well in the millions of diesel vehicles already operating on roads, rails, and waterways around the world and can be stored in existing tanks at local filling stations and fuel dealerships. Biodiesel is unquestionably a good first step toward greater energy diversification and independence.

Gene Gebolys says that he used to agree with the bridge technology view, but now he's not so sure. "I don't know if it's a bridge technology or not," he says. "I just know that it's so much better than the alternative of doing nothing. What the future holds is anybody's guess. I think that the days of predicting that we are going to run out of oil in thirty years are over. We don't know exactly when we are going to run out of oil; we just know that the amount of oil is finite and that we are eventually going to run out. We don't know how much human activity is leading to global warming, but we do know that we are affecting that process with a reasonable amount of certainty. We also know, in general, that our reliance on foreign oil leads to instability in the Middle East. We know, in general, that these aren't good things, and therefore we are going to have to do whatever we can to begin to address them. I think biodiesel is a relatively painless way to do that, and I think it's going to be around for a long time. Whether it's a bridge to something else, I don't know. It may well be, but that's for the next generation to figure out. This generation should be focused on getting biodiesel into the mainstream."[30]

Daryl Reece of Pacific Biodiesel agrees about the long-term prospects for biodiesel. "I don't see the diesel engine being replaced anytime soon," he says. "I think we're going to have it around for many, many years in order to produce the torque and horsepower needed to run the many different types of equipment that rely on it. Gasoline-powered engines just can't match that kind of performance. That means there are good markets and excellent opportunities for biodiesel."[31]

The biodiesel industry is still young and in many ways like a recent college graduate. After years of study and hard work, the industry is finally embarked on a great adventure. And like any young person, the industry faces many challenges ahead. It needs to settle the feedstock wars and develop better feedstock choices to help level out the worst fluctuations in the commodities market. The industry needs to learn to work more cooperatively with the many different constituencies—both external and internal—that can help it succeed. Large producers, small producers, and home-brew advocates need to be allowed, and even helped, to find and develop their respective market niches. And all producers, regardless of their size, need to focus even more attention on producing a top-quality product—all of the time. The industry also needs to find the right balance between receiving government support and avoiding becoming totally dependent on it. It needs to take the steady, long-term approach of developing a stable, dependable, and ultimately profitable segment of the energy sector. The industry is now a global reality, and it faces global challenges. But those challenges can be met, and ultimately overcome, with greater international cooperation. Finally, the biodiesel industry needs to find—and settle into—its proper place in the new renewable energy economy of the twenty-first century that is being born even as the old fossil fuel economy comes to an end. Rudolf Diesel would be pleased.

ORGANIZATIONS AND ONLINE RESOURCES

Australia

Australian Biofuels Association
P.O. Box 3753
Manuka, ACT 2603
Phone: 61-2-6295-2399
An industry group that provides useful information on biodiesel, ethanol, and more.

Biodiesel Association of Australia
P.O. Box 301
Homebush South, NSW 2140
Phone: 61-2-9746-7617
Web site: www.biodiesel.org.au
An industry association whose Web site offers a wide range of information on biodiesel, including facts, standards, news, an online library, links, and much more.

Austria

Austrian Biofuels Institute
Graben 14/2, Pf. 97
A-1014 Vienna
ATU4168608
Phone: 43-1-534-56-33
Web site: www.biodiesel.at
An international center of expertise on liquid biofuels (in German and English).

Canada

Canadian Renewable Fuels Association
Web site: www.greenfuels.org
An industry-association Web site that offers information on ethanol and biodiesel with links and more (in English and French).

Europe

European Biodiesel Board
Brd Saint-Michel 47
1040 Brussels, Belgium
Phone: 32-2-737-76-13
E-mail: ebb@ebc-youroffice.com
Web site: www.ebb-eu.org
A nonprofit organization with the aim of promoting the use of biodiesel in the European Union.

Teaching Chemistry By Vegetable Oil Theme
Web site: http://koal2.cop.fi/leonardo/
The Web site of a European vocational education chemistry program that focuses on biodiesel and vegetable oil chemistry including lab safety information and an extensive analytical section (some in German).

France

Club des Villes Diester France
Web site: www.villesdiester.asso.fr
An educational and collaborative network of biodiesel users, producers, and institutions, formed to help promote greater use of biodiesel (in French and some English).

Germany

Union for the Promotion of Oil and Protein Plants (Union zur Förderung von Oel- und Proteinpflanzen, or UFOP)
E-mail: ufop@wpr-communication.de
Web site: www.ufop.de
An educational and promotional Web site containing a wide range of resources with both a German and international focus (in German and English).

Japan

Nanohana Project Network
1273-5 Kamitoyoura, Azuchicho, Gamou-gun, Shiga Pref.
Shiga Prefecture Environment and Consumer Cooperative
Phone: 81-748-46-4551
Fax: 81-748-46-4550
E-mail: nanohana@nanohana.gr.jp
An organization dedicated to the local production and consumption of biodiesel from used cooking oil made from locally grown agricultural feedstocks.

United Kingdom

Allied Biodiesel Industries (UK)
Web site: www.biofuels.fsnet.co.uk/biobiz.htm
An organization representing the British biodiesel industry; the Web site offers information and links to suppliers, filling station locations, and more.

British Association for Bio Fuels and Oils
E-mail: info@biodiesel.co.uk
Web site: www.biodiesel.co.uk
An organization dedicated to the promotion of transport fuels and oils from renewable sources. The Web site offers a good selection of resources and links, information on emissions, and more.

United States

Alternative Fuels Data Center
Web site: www.eere.energy.gov/cleancities/afdc/
A comprehensive site for information on alternative fuels and vehicles.

National Biodiesel Board
P.O. Box 104888
Jefferson City, MO 65110-4898
Phone: 800-841-5849
Web site: www.biodiesel.org
The Web site is an excellent and extensive source of current industry information on biodiesel, news, technical information, links, and much more.

National Renewable Energy Laboratory (NREL)
1617 Cole Boulevard
Golden, CO 80401
Phone: 303-275-3000
E-mail: webmaster@nrel.gov
Web site: www.nrel.gov
The leading center for renewable energy research in the United States.

Cooperatives (U.S.)
The Berkeley Biodiesel Collective
Berkeley, California
E-mail: berkeleybiodiesel@yahoo.com

The Biofuels Research Cooperative (Straight Vegetable Oil)
Sebastopol, California
E-mail: veggieoilcoop@yahoo.com
Web site: www.veggieoilcoop.org

Boulder Biodiesel Cooperative
Boulder, Colorado
Phone: 303-449-3277
Web site: www.boulderbiodiesel.com

The Grease Works! Biodiesel Cooperative
Corvallis, Oregon
Email: justin@grease-works.com
Web site: www.grease-works.com

Pioneer Valley Biodiesel Cooperative
Greenfield, Massachusetts
Phone: 413-247-0163
E-mail: biodieselcoop@aol.com

Portland's Biodiesel Cooperative
Portland, Oregon
E-mail: info@gobiodiesel.com
Web site: www.gobiodiesel.com

Other Biodiesel Resources
Biodiesel America
Web site: www.biodieselamerica.org
Comprehensive Web site about small-scale biodiesel production and more. Home of the Veggie Van, offering comprehensive information about biodiesel, the popular book on how to make it yourself, biodiesel history, FAQs, products, and much more.

Distribution Drive
Biodiesel links
Web site: http://www.distributiondrive.com/links.html
An exhaustive list of biodiesel and industry links on the site of a Texas biodiesel supplier.

Fat of the Land
Web site: www.lardcar.com
Web site for the classic 1995 film Fat of the Land, *describing the humorous adventures of five women who drove across the United States in a biodiesel-powered van.*

Journey to Forever
Web site: journeytoforever.org
A highly educational Web site containing a wealth of information about biodiesel and many other sustainable-living subjects, including links to additional sources (in English, Japanese, and Chinese).

Veggie Avenger
Web site: www.veggieavenger.com
Comprehensive information about biofuels and related technology, with news, glossary, numerous links, and much more about biodiesel and straight vegetable oil.

Online Discussion Groups

Biodiesel Discussion Forum
Web site: biodiesel.infopop.cc
A discussion forum and message board sponsored by the BioBeetle, Maui Recycling Service, and Pacific Biodiesel.

BiodieselNow.com
Web site: www.biodieselnow.com
A comprehensive site that offers answers to virtually any question about biodiesel in the United States and around the world (foreign-language skills helpful for some message threads). Considered by many to be the *biodiesel online discussion site.*

NOTES

Introduction

1. George Monbiot, "Bottom of the Barrel," *The Guardian*, December 2, 2003.

Chapter 1: Rudolf Diesel

1. W. Robert Nitske and Charles Morrow Wilson, *Rudolf Diesel: Pioneer of the Age of Power* (Norman, Okla.: University of Oklahoma Press, 1965), 15–17. Unless noted otherwise, most of the information on Rudolf Diesel that follows is from the same source.
2. In a later account of this first engine test given in a speech by Diesel on April 13, 1912, in Saint Louis, Missouri, the first engine was said to be fueled by powdered coal dust and the explosion nearly fatal. This later account is at odds with earlier records of the experiment.
3. Walter Kaiser, "Rudolf Diesel and the Second Law of Thermodynamics," *German News Magazine*, June/July 1997. http://www.germanembassy-india.org/news/june97/76gn16.htm. (This monthly magazine is published by the German embassy of India in New Delhi.)
4. According to most sources, Diesel did not begin to experiment with coal dust as a fuel for his engine until 1897.
5. Gerhard Knothe, "Historical Perspectives on Vegetable Oil–Based Diesel Fuels," *Inform: International News on Fats, Oils and Related Materials* 12 (November 2001): 1104. (*Inform* is a monthly publication of the American Oil Chemist's Society in Champaign, Illinois.)

Chapter 2: Vegetable Oil Revival

1. W. Robert Nitske and Charles Morrow Wilson, *Rudolf Diesel: Pioneer of the Age of Power* (Norman, Okla.: University of Oklahoma Press, 1965), 214.
2. Ibid., 213.
3. Ibid., 208.
4. Ibid., 259.
5. Gerhard Knothe, "Historical Perspectives on Vegetable Oil–Based Diesel Fuels," *Inform: International News on Fats, Oils and Related Materials* 12 (November 2001): 1106.
6. Darryl Melrose, telephone interview by the author, March 23, 2004.
7. Knothe, "Historical Perspectives," 1105.
8. Ibid.
9. Manfred Wörgetter, e-mail interview by the author, January 14, 2004.
10. Knothe, 1107.
11. Martin Mittelbach, telephone interview by the author, January 15, 2004.
12. Manfred Wörgetter, e-mail to the author, May 27, 2004.

13. Werner Körbitz, "The Biodiesel Market Today and Its Future Potential," in *Proceedings of the Plant Oils as Fuels—Present State of Science and Future Developments Symposium* (held in Potsdam, Germany, February 16–18, 1997), (Berlin: Springer-Verlag, 1998), 4.
14. Werner Körbitz, telephone interview by the author, January 12, 2004.
15. Lourens du Plessis, telephone interview by the author, February 2, 2004.
16. Charles Peterson, telephone interview by the author, January 12, 2004.
17. Ibid.
18. Ibid.
19. H. E. Haines and J. Evanoff, "Environmental and Regulatory Benefits Derived from the Truck in the Park Biodiesel Emissions Testing and Demonstration in Yellowstone National Park" (paper presented at the Bioenergy '98 Conference of the U.S. Department of Energy Regional Bioenergy Program in Madison, Wisconsin, October 1998; paper revised December 8, 1998).

Chapter 3: Biodiesel 101

1. Werner Körbitz, telephone interview by the author, January 12, 2004.
2. All yield figures below are from the chart "Vegetable Oil Yields" located at http://journeytoforever.org/biodiesel_yield.html.
3. James A. Duke, *Handbook of Energy Crops* (1983), an electronic publication on NewCROP, the New Crop Resource Online Program hosted by the Purdue University Center for New Crops & Plant Products Web site, located at http://www.hort.purdue.edu/newcrop/duke_energy/dukeindex.html. Some of the oil-crop information that follows is based on the same source.
4. Werner Körbitz, *New Trends in Developing Biodiesel World-wide* (Vienna: Austrian Biofuels Institute, 1999). All subsequent percentages for global biodiesel raw material sources are from a chart contained in this report.
5. John Sheehan et al., *A Look Back at the U.S. Department of Energy's Aquatic Species Program: Biodiesel from Algae* (Golden, Colo.: National Renewable Energy Laboratory, July 1998), ii.
6. Ibid., iii.
7. J. Connemann and J. Fischer, "Biodiesel in Europe 2000: Biodiesel Processing Technologies and Future Market Development" (paper presented at the symposium Biodiesel—Fuel from Vegetable Oils for Compression-Ignition Engines at the Technische Akademie Esslingen, May 17, 1999, Ostfildern/Stuttgart, Germany).
8. Some recent performance results in school bus fleets have indicated an increase in fuel efficiency, so the jury is still out on this performance factor.
9. John Sheehan et al., *Life Cycle Inventory of Biodiesel and Petroleum Diesel for Use in an Urban Bus* (Golden, Colo.: National Renewable Energy Laboratory, May 1998).
10. Ibid.

11. Hosein Shapouri, James A. Duffield, and Michael Wang, *The Energy Balance of Corn Ethanol: An Update*, Agricultural Economic Report No. 814 (Washington, D.C.: U.S. Department of Agriculture, Office of the Chief Economist, Office of Energy Policy and New Uses, July 2002).
12. Werner Körbitz, *The Technical, Energy and Environmental Properties of BioDiesel* (Vienna: Körbitz Consulting, 1993).

Chapter 4: Biodiesel's Many Uses

1. Onno Syassen, "Diesel Engine Technologies for Raw and Transesterified Plant Oils as Fuels: Desired Future Qualities of the Fuels," in *Proceedings of the Plant Oils as Fuels—Present State of Science and Future Developments Symposium* (held in Potsdam, Germany, February 16–18, 1997), (Berlin: Springer-Verlag, 1998), 52.
2. Bosch, "At High Pressure: 75 years of Bosch Diesel Injection," http://archive.bosch.com/en/archive/theme_11_2002.htm.
3. National Biodiesel Board, "Lambert International Airport," http://www.biodiesel.org/resources/users/stories/lambert.shtm.
4. Bernhard Prossnigg, "Experiences with Biodiesel in the Bus Fleet of the Public Transportation System of the City of Graz" (a lecture presented at From the Frying Pan into the Tank: Recycled Frying Oil Collection and Its Use as Biodiesel in Styria, a seminar in Graz, June 29, 2000).
5. National Biodiesel Board, "School Buses," http://www.nbb.org/markets/sch/default.asp.
6. National Biodiesel Board, "Medford, New Jersey School District," http://www.biodiesel.org/resources/users/stories/medfordnj.shtm.
7. National Biodiesel Board, "Farmer Use," http://www.nbb.org/pdf_files/farmer_use.pdf.
8. National Biodiesel Board, "Easier on Marine Environment," http://www.nbb.org/markets/mar/default.asp.
9. Mario Osava, "Biodiesel Trains on the Right Track," Inter Press Service News Agency Web site, http://ipsnews.net/africa/interna.asp?idnews=21707.
10. Denver Lopp and Dave Stanley, *Soy-Diesel Blends Use in Aviation Turbine Engines* (West Lafayette, Ind.: Purdue University, Aviation Technology Department, 1995).
11. "CytoSol, Exxon Demonstrate Solvent Capabilities in Texas," *Feedstocks: News about Industrial Products Made from Soy* 4, no. 3 (1999): 1, 2.
12. John Van de Vaarst, telephone interviews by the author, May 6, 2003, and June 27, 2003.
13. Jenna Higgins, telephone interview by the author, July 3, 2003.
14. Bob Cerio, telephone interview by the author, June 27, 2003.
15. Ralph Mills, telephone interview by the author, July 9, 2003.
16. Joel Glatz, telephone interview by the author, July 4, 2003.
17. Stephan Chase, telephone interview by the author, July 7, 2003.
18. National Biodiesel Board press release, February 9, 2004.

Chapter 5: Europe, the Global Leader

1. USDA, "EU: Biodiesel Industry Expanding Use of Oilseeds," September 20, 2003, Production Estimates and Crop Assessment Division, Foreign Agricultural Service, U.S. Department of Agriculture, 1, 2. http://www.fas.usda.gov/pecad2/highlights/2003/09/biodiesel3/index.htm.
2. The EU-15 members (Austria, Belgium, Denmark, Finland, France, Germany, Greece, Ireland, Italy, Luxembourg, Portugal, Spain, Sweden, the Netherlands, and the United Kingdom) plus new members Cyprus, the Czech Republic, Estonia, Hungary, Latvia, Lithuania, Malta, Poland, Slovakia, and Slovenia.
3. Used frying oil is generally called waste vegetable oil (WVO) in the United States, but the different terms refer to the same commodity.
4. Raffaello Garofalo, telephone interview by the author, February 24, 2004.
5. The percentage represented by feedstock cost can vary from about 60 percent to as much as 80 percent, depending on the feedstock and the production technology used.
6. *Biofuels: Emerging Developments and Existing Opportunities* (New York: Technical Insights Inc., 2002). For a summary, go to the Web site, http://www.the-infoshop.com/study/ti12890_biofuels.html.
7. Werner Körbitz, telephone interview by the author, January 12, 2004.
8. Werner Körbitz and Jens Kossmann, "Production and Use of Biodiesel," in *New and Emerging Bioenergy Technologies*, Risø Energy Report 2, ed. Hans Larsen, Jens Kossmann, and Leif Sønderberg (Roskilde, Denmark: Risø National Laboratory, November 2003), 31.
9. Garofalo, interview.
10. USDA, *EU: Biodiesel Industry Expanding Use of Oilseeds*, 5.
11. Garofalo, interview.
12. Ibid.
13. Energy and Transport Directorate-General of the European Commission, "Energy Taxation: Commission Proposes Transitional Periods for Accession Countries," http://europa.eu.int/comm/energy_transport/mm_dg/newsletter/nl80-2004-01-30_en.html#EN%2001.
14. Austrian Biofuels Institute, *Biodiesel—A Success Story: The Development of Biodiesel in Germany*, a report for the International Energy Agency, Bioenergy Task 27, Liquid Biofuels (Vienna: Austrian Biofuels Institute, June 2001, update February 2002), 21.
15. Most of the recent German biodiesel plant expansion information is based on the chart "Biodieselproduktionskapazitaten in Deutschland," which is located at the Web site http://www.iwr.de/biodiesel/kapazitaeten.html.
16. UFOP, *Biodiesel in Bus Fleets: The Kreiswerke Heinsberg's GmbH and Stadtwerke Neuwied's Experience* (Bonn: UFOP), 1–34.
17. Hans Plaettner-Hochwarth and Klaus Schreiner, *Biodiesel und Sportschiffart in der Eurigio Bodensee* (Bonn: UFOP, 2001), 1–33.
18. Dieter Bockey, *Biodiesel Production and Marketing in Germany: The Situation and Perspective* (Berlin: UFOP, 2002), 8.

19. Austrian Biofuels Institute, *Biodiesel—A Success Story*, 14–17.
20. Garofalo, interview.
21. France is second only to Spain in EU ethanol production, but ethanol is generally not as important as biodiesel in the EU due to low corn production and a higher proportion of diesel engines compared to the United States.
22. *Biodiesel: Documentation of the World-Wide Status 1997*, a report for the International Energy Agency (IEA), commissioned by the BLT-Federal Institute for Agricultural Engineering (Wieselburg, Austria: Austrian Biofuels Institute, 1997), 20.
23. Energy Systems Research Unit, "Biofuels in Europe" and "EU Policy," http://www.esru.strath.ac.uk/EandE/Web_sites/02-03/biofuels/foreign_europe.htm.
24. Marie-Cécile Hénard and Xavier Audran, *France, Agricultural Situation, French Biofuel Situation*, Global Agricultural Information Network report FR3044 (Washington, D.C.: U.S. Department of Agriculture, Foreign Agricultural Service, 2003), 3.
25. Ibid., 4.
26. Ibid., 6.
27. Ibid., 4.

Chapter 6: Other European Countries

1. *Biodiesel: Documentation of the World-Wide Status 1997*, a report prepared for the International Energy Agency (IEA), commissioned by the BLT-Federal Institute for Agricultural Engineering (Wieselburg, Austria: Austrian Biofuels Institute, 1997), 23.
2. Liquid Biofuels Network, *Liquid Biofuels Activity Report* (France: EUBIONET, Liquid Biofuels Network, April 2003), 32.
3. Manfred Wörgetter, e-mail to the author, May 27, 2004.
4. Bernhard Prossnigg, "Experiences with Biodiesel in the Bus Fleet of the Public Transportation System of the City of Graz" (a lecture presented at From the Frying Pan into the Tank: Recycled Frying Oil Collection and Its Use as Biodiesel in Styria, a seminar in Graz, June 29, 2000).
5. Z. Sedivá and P. Jevi, *Present State of Production and Sale of Biofuels on Base of Rape Oil in the Czech Republic* (Prague: Research Institute of Agricultural Engineering, Association for Biodiesel Production, 2001), 117.
6. *Biodiesel: Documentation of the World-Wide Status 1997*, 25, 26.
7. Martin Cvengros and Ján Cvengros, "Review on Development and Legislation of Biodiesel Production and Utilization in Slovakia" (a paper presented at the Techagro Fair in Brno, the Czech Republic, April 2002), 3.
8. Ibid., 7, 8.
9. Spanish Ministry of Economics, http://www2.mineco.es/Mineco/Comunicacion/Noticias/RATO+BIODIESEL.htm (page now removed from Web site).

10. Austrian Biofuels Institute, *Annual Report 2002* (Vienna: Austrian Biofuels Institute, 2002), 4.
11. Spanish Ministry of Economics, "La planta de producción de biodiesel del IDAE en Alcalá de Henares."
12. P. H. Mensier, "L'emploi des Huiles Végétales Comme Combustible dans les Moteurs," *Oléagineux*, Février (1952), 69.
13. European Energy Crops InterNetwork, "Evolution of Rape in Belgium and Its Utilization as Biofuel," document ID B10082, November 7, 1997, http://btgs1.ct.utwente.nl/eeci/archive/biobase/B10082.html (Web site now discontinued).
14. The Netherlands Agency for Energy and the Environment, "The Netherlands—ATEP Builds Biodiesel Factory in Arnhem," http://www.novem.nl/default.asp?documentId=21376.
15. Comments posted on the Farming Life Web site (http://www.farminglife.com) under the heading, "United Kingdom: Biofuel Blow," on December 15, 2003 (Web page has expired).
16. Reuters UK, "Tesco to Sell Rapeseed Biodiesel," May 3, 2004, http://www.reuters.co.uk/newsPackageArticle.jhtml?type=topNews&storyID=502389§ion=news (Web page has expired).
17. Green Shop, "The First UK Garage to Sell Biodiesel," http://www.greenshop.co.uk/news/Biodiesel-Opening-2002.htm.
18. Karen McLauchlan, "Petroplus Fuelling Biodiesel Drive," *The Evening Gazette*, January 22, 2004, http://icteesside.icnetwork.co.uk/0400business/0004tod/page.cfm?objectid=13845660&method=full&siteid=50080.
19. Greenergy press release, April 14, 2004.
20. Liquid Biofuels Network, *Liquid Biofuels Network Activity Report*, 30.
21. Dr. Sigitas Lazauskas, Lithuanian Institute of Agriculture, "Non-food Crop Activity in the Baltic Sea Region," published in the *Interactive European Network for Industrial Crops and their Applications*, Newsletter Number 21, December 2003, http://www.ienica.net/newsletters/newsletter21.pdf.
22. "What Is Biodiesel?" http://www.betancalibration.com/pdf/BioDiesel.pdf.
23. Baltic News Service, March 6, 2003 (http://terminal.bns.lv/login.jsp) (Web page has expired).
24. Forum Latvia, http://www.forumpress.it/latvia0/p14.htm (accessed March 11, 2004).
25. "Polish Biodiesel," Biodiesel.pl, http://www.biodiesel.pl (Web page has expired).
26. *Puls Biznesu*, December 3, 2003, 11.
27. Rico Cruz, phone interview by the author, April 13, 2004.

Chapter 7: Non-European Countries

1. *Investigation into the Role of Biodiesel in South Africa*, a report to the Department of Science and Technology (Pretoria: CSIR Transportek, March 2003).

2. UNIDO, *CDM Investor Guide*, South Africa (Vienna: United Nations Industrial Development Organization, February 2003), 29.
3. Earthlife Africa/WWF, *Employment Potential of Renewable Energy in South Africa* (Johannesburg: Earthlife Africa / Denmark: WWF, November 2003), 45.
4. "A Pioneering Bio-Fuel Project Holds Out Unprecedented Empowerment Opportunities," *Echo,* supplement to *The Natal Witness*, Thursday, July 29, 2003.
5. Darryl Melrose, telephone interview by the author, March 23, 2004.
6. Justin Brown, *Sasol Looking at Making Soyabean Biodiesel* (Pretoria: Department of Trade and Industry, March 11, 2004).
7. Elnette Oelofse, "Four Little Seeds Helping to Empower Africa," September 2002, the International Oracle Syndicate, http://www.oraclesyndicate.org/pub_e/e.oelofse/pub_9-02_2.htm#top.
8. Good News India, "Rising Bio-diesel Tide," September 18, 2003, http://www.goodnewsindia.com/Pages/content/updates/story/117_0_4_0_C/.
9. Good News India, "Honge Oil Proves to Be a Good Biodiesel," http://www.goodnewsindia.com/Pages/content/discovery/honge.html.
10. Ibid.
11. S. Srinivasan, "Isolated Hamlet in Indian Forest Gets Electricity from Seed-Powered Generator," Associated Press, October 15, 2003.
12. RenewingIndia.org, "Biodiesel: First Trial Run on Train," http://www.renewingindia.org/news1_jan_biodiesel.html.
13. On the DaimlerChrysler Web site, http://www.daimlerchrysler.com/, click on **Environment** and look for the article "Start-up of Jatropha Biodiesel Production in India."
14. The Hindu Business Line, "IOC to Start Field Trials of Biodiesel," December 10, 2003, http://www.thehindubusinessline.com/businessline/blnus/14101710.htm.
15. "Five Biodiesel Plants Sanctioned in State," *The Central Chronicle*, December 31, 2003.
16. "Thai Navy Joins Biodiesel Feasibility Study," *Bangkok Post*, May 13, 2003.
17. *Biodiesel: Documentation of the World-Wide Status 1997*, a report prepared for the International Energy Agency (IEA), commissioned by the BLT-Federal Institute for Agricultural Engineering (Wieselburg, Austria: Austrian Biofuels Institute, 1997), 18.
18. Austrian Biofuels Institute, *Annual Report 2002* (Vienna: Austrian Biofuels Institute, 2002), 7.
19. Journey to Forever, http://journeytoforever.org/.
20. "About the Company," Pacific Biodiesel, http:// www.biodiesel.com/aboutPacBio.htm.
21. Yasuji Nagai, "Fields Bloom with New Eco-Friendly Fuel Alternative," *The Asahi Shimbun*, April 29, 2004, http://www.asahi.com/english/lifestyle/TKY200404290137.html.

22. National Biodiesel Board, "Japan Creates First Co-Generation Turbine Fueled by Biodiesel," *Biodiesel Bulletin, A Monthly Newsletter of the National Biodiesel Board,* http://www.biodiesel.org/news/bulletin/2002/072002.pdf.
23. "Government Unveils Biomass-Fuel Project," *The Japan Times,* December 28, 2002, http://www.japantimes.co.jp/cgi-bin/getarticle.pl5?nb20021228a9.htm.
24. Japan for Sustainability, "Making Biodiesel Fuel from Sunflower Oil," http://www.japanfs.org/db/database.cgi?cmd=dp&num=484&UserNum=&Pass =&AdminPa.
25. Nanohana Project Network, http://www.eic.or.jp/jfge/english/projects/P13.html (in Japanese).
26. Department of Environment and Natural Resources, "'Biodiesel' to the Public," http://denr.gov.ph/article/articleprint/628/-1/152/.
27. *Biodiesel: Documentation of the World-Wide Status 1997*, 24.
28. Abdul Khalik, "Biodiesel Fuel Could Free Jakarta from Pollution," *The Jakarta Post*, January 9, 2004, http://www.ecologyasia.com/NewsArchives/jan2004/ thejakartapost.com_20040109_2.htm.
29. "PNG to Develop Coconut Oil As Biodiesel Fuel," TheNational.com, November 14, 2003.
30. Biodiesel Industries, Las Vegas, Nevada, press release, March 13, 2003.
31. Amadeus Energy Limited, "Biodiesel Project," http://www.amadeusenergy .com/default.asp?V_DOC_ID=841.
32. Suzi Kerr, Brian White, et al., *Renewable Energy and the Efficient Implementation of New Zealand's Current and Potential Future Greenhouse Gas Commitments* (New Zealand: East Harbour Management Services Limited, August 14, 2002).
33. Massey University, "Meridian Energy, Massey Investigate Biodiesel Option," June 9, 2003, http://masseynews.massey.ac.nz/2003/masseynews/june/june16/ stories/fuel.html.
34. "Brazil Biodiesel—Soy Oil Stocks," *Queensland Independent*, December 6, 2001.
35. Alexander's Gas & Oil Connections, "Latin America Is Turning Clean and Green," http://www.gasandoil.com/goc/history/welcome.html (click on **Latin America**, then look for the article in the list that appears).
36. *Renewable Energy: State of the Industry Report* 9, April–June 2003 (Morrilton, Ark.: Winrock International, 2003), 7, 9.
37. Mario Osava, "Biodiesel Trains on the Right Track," Inter Press Service News Agency, http://ipsnews.net/africa/interna.asp?idnews=21707.
38. Carlos Büttner, "Biodiesel: Characterización, Procesos de Elaboración, Y Normas de Control," http://www.mic.gov.py/combustibles/Presentacin_Bttner.ppt.
39. Carrie Gibson, "CU Biodiesel Welcome New Production Factory," April 8, 2004, http://bcn.boulder.co.us/campuspress/messages/1732.html.
40. Proyecto Biomasa, "Biomass Project Nicaragua," http://www.ibw.com.ni/~biomasa/.
41. Tierramérica, "Costa Rica: Promoting Biodiesel," March 26, 2004, http://www .tierramerica.net/2004/0105/iecobreves.shtml.

42. City of Brampton, Ontario, "Brampton Transit Powered by Biodiesel," press release, October 24, 2003, http://www.city.brampton.on.ca/press/03-197.tml.
43. National Biodiesel Board, press release, http://www.biodiesel.org/resources/memberreleases/20040302_TOPIA_CANADAS_FIRST_BIODIESEL_PUMP.pdf.
44. Ian Ross, "Biodiesel Plant First Anchor Tenant," Northern Ontario Business, May 2003, http://www.northernontariobusiness.com/displayHeadline.asp?165id115-pn=&view=22190.

Chapter 8: A Brief History

1. Bill Kovarik, "Henry Ford, Charles Kettering, and the 'Fuel of the Future,'" *Automotive History Review*, no. 32 (Spring 1998), 7–27, reproduced on the Web at http://www.radford.edu/~wkovarik/papers/fuel.html.
2. European Biofuels Group, "A History of Biodiesel/Biofuels," http://www.eurobg.com/biodiesel_history.html (Web site now discontinued).
3. Thomas Reed, telephone interview by the author, April 6, 2004.
4. Ibid., as well as information from the biodiesel page on Reed's Web site at http://www.woodgas.com/biodies.htm.
5. Leon Schumacher, telephone interview by the author, April 2, 2004.
6. Ibid.
7. Bill Ayres, telephone interview by the author, April 9, 2004.
8. Ibid.
9. "Florida Kids and Keys Benefit from Biodiesel," *Biofuels Update: Report on U.S. Department of Energy Biofuels Technology* 4, no. 3 (Fall 1996): 1, 3.
10. Gary Haer, telephone interview by the author, April 14, 2004.
11. "Biodiesel Plants Sprout Up Across the United States," *Biofuels Update: Report on U.S. Department of Energy Biofuels Technology* 5, no. 1 (Winter 1997): 1, 3, 6.
12. Joe Loveshe, e-mails to the author, April 12, 2004.
13. Bob King, telephone interview by the author, May 4, 2004.
14. Pacific Biodiesel, http://biodiesel.com/company_profile.htm (Web page now discontinued; the new profile is located at www.biodiesel.com/aboutPacBio.htm).
15. King, interview.
16. Neil Caskey, telephone interview by the author, April 6, 2004.
17. National Biodiesel Board, "Who Are We?" http://www.nbb.org/aboutnbb/whoarewe/.
18. Joe Jobe, telephone interview by the author, May 11, 2004.
19. Haer, interview.
20. Nicole Cousino, telephone interview by the author, April 10, 2004.
21. Ibid.
22. Sarah Lewison, telephone interview by the author, April 11, 2004.
23. The Veggie Van, "Joshua Tickell Bio," http://www.joshuatickell.com/bio/index.php.
24. Cousino, interview.

Chapter 9: The Main Players

1. Jon Van Gerpen, telephone interview by the author, April 12, 2004.
2. Mark Zappi, Rafael Hernandez, et al., A *Review of the Engineering Aspects of the Biodiesel Industry*, a report for the Mississippi Biomass Council (Jackson, Miss.: Mississippi University Consortium for the Utilization of Biomass, Mississippi State University, August 2003), 1, 8, 19.
3. Dennis Griffin, telephone interview by the author, April 19, 2004.
4. Ibid.
5. Gene Gebolys, telephone interview by the author, April 14, 2004.
6. Ibid.
7. Bob Clark, telephone interview by the author, April 19, 2004.
8. Ibid.
9. Business Wire, "Bakersfield Biodiesel Plant Completes First Week of Production Testing While Congress Debates Final Energy Bill," November 6, 2003, http://www.greenstarusa.com/news/03-11-06.html.
10. All information about American Biofuels comes from company press releases located at http://www.greenstarusa.com/news/press.html.
11. Jerrel Branson, e-mail to the author, April 12, 2004.
12. Jerrel Branson, telephone interview by the author, April 19, 2004.
13. Jeff Probst, telephone interview by the author, May 6, 2004.
14. Ibid.
15. John Hurley, telephone interview by the author, April 29, 2004.
16. Ibid.
17. Maria "Mark" Alovert, "The Grease Trap: Co-ops part 1," May 21, 2003, http://lists.subtend.net/pipermail/pdx-biodiesel/2003-May/000721.html.
18. Tom Leue, telephone interview by the author, April 30, 2004.
19. Jerrel Branson, telephone interview by the author, November 13, 2003.
20. Joe Jobe, telephone interview by the author, May 11, 2003.

Chapter 10: Biodiesel Politics

1. Neil Caskey, telephone interview by the author, April 6, 2004.
2. Charles Hatcher, telephone interview by the author, April 23, 2004
3. National Biodiesel Board, "Biodiesel Poised to be a Significant Contributor to the U.S. Alternative Fuels Market," January 2000, http://www.biodiesel.org/resources/reportsdatabase/reports/gen/20000102_gen-212.pdf.
4. Hatcher, interview.
5. Daryl Reece, telephone interview by the author, November 24, 2003.
6. H. Josef Hebert, "EPA Targets Off-Road Vehicles, Marine Vessels in New Pollution Controls," Associated Press, May 12, 2004.
7. A filibuster is a procedural tactic in the U.S. Senate whereby a senator can block a bill by talking continuously until the bill is either withdrawn or the bill's supporters can muster sixty votes to invoke "cloture," cutting off the filibuster and moving the bill forward.

8. National Biodiesel Board, "Senate Passes Jobs Bill Including Biodiesel Tax Provisions," press release, May 12, 2004.
9. Caskey, interview.
10. Dennis Griffin, telephone interview by the author, April 19, 2004.
11. Bill Ayres, telephone interview by the author, April 9, 2004.
12. Jon Van Gerpen, telephone interview by the author, April 12, 2004.
13. Hatcher, interview.
14. Ibid.
15. Gene Gebolys, interview by the author, April 14, 2004.
16. Hatcher, interview.
17. Ibid.
18. National Biodiesel Board, press release, March 15, 2002.
19. National Biodiesel Board, press release, April 4, 2004.
20. National Biodiesel Board, press release, June 12, 2003.
21. Hatcher, interview.
22. Joe Jobe, telephone interview by the author, May 11, 2004.
23. Bob King, telephone interview by the author, May 4, 2004.
24. Ibid.
25. Rico Cruz, telephone interview by the author, April 13, 2004.
26. Jobe, interview.
27. Reece, interview.

Chapter 11: Recent Developments

1. National Biodiesel Board, "Bonnie Raitt Fuels Up with Cleaner Burning Biodiesel on Tour," press release, September 17, 2002.
2. National Biodiesel Board, "Indigo Girls Use Cleaner Burning Biodiesel on Tour," press release, April 14, 2003.
3. National Biodiesel Board, "Neil Young Goes on Tour with Cleaner Burning Biodiesel," press release, February 20, 2004.
4. National Biodiesel Board, "Biodiesel Awards Recognize Daryl Hannah, Industry Leaders," press release, February 2, 2004.
5. Biodiesel Industries Inc., "Biodiesel Industries Opens First Biodiesel Filling Station in US," http://www.pipeline.to/biodiesel/ (Web page expired).
6. San Joaquin Valley Clean Cities Coalition, "First Public Biodiesel Fueling Station to Open in San Francisco on May 23, 2001," May 23, 2001, http://www.valleycleancities.org/Articles/05162001D.html.
7. SolarAccess, "Biodiesel Power for Colorado School Buses," November 7, 2003, http://www.solaraccess.com/news/story?storyid=5489.
8. Gail R. Frahm, "St. Johns School Buses Rolling 1 Million Miles on Biodiesel," Michigan Soybean Promotion Committee press release, April 27, 2004.
9. It should be noted that some earlier fleet tests have noted a slight decrease in fuel efficiency with the use of biodiesel, so the jury is still out on this issue.
10. National Biodiesel Board, "Back to School with Biodiesel," press release, October 6, 2003.

11. American Public Transportation Association, "Transit Resource Guide," no. 5, rev. April 2004, http://www.apta.com/research/info/briefings/briefing_5.cfm.
12. Clean Cities Program (U.S. Department of Energy), "Alternative Fuel Success Stories, Five Seasons Transportation and Parking," http://www.ccities.doe.gov/success/five_seasons.shtml.
13. Clean Cities Program (U.S. Department of Energy), "Alternative Fuel Success Stories, Bi-State Transit Agency," http://www.eere.energy.gov/cleancities/progs/new_success_ddown.cgi?25.
14. National Biodiesel Board, "U.S. Military Facilities Increasingly Fill Up With Biodiesel," press release, June 16, 2003, http://www.biodiesel.org/resources/pressreleases/fle/20030616_military_users.pdf.
15. Ibid.
16. National Biodiesel Board, "U.S. Navy to Produce Its Own Biodiesel," press release, October 30, 2003, http://www.nbb.org/resources/pressreleases/gen/20031030_navy_to_produce_biodiesel.pdf.
17. Ibid.
18. The Soy Daily, "Biodiesel Pump Opens to the Public at Yellowstone," October 15, 2002, http://www.thesoydailyclub.com/BiodieselBiobased/yellowstone10182002.asp.
19. Kentucky Soybean News, "Mammoth Cave Becoming an Environmental Leader," http://www.kysoy.org/news/mammothcave.htm.
20. National Biodiesel Board, "Biodiesel Users: Channel Islands National Park," http://www.biodiesel.org/resources/users/stories/channelisle.shtm.
21. "Yosemite to Produce Its Own Biodiesel," *The Biodiesel Bulletin*, December 1, 2003.
22. National Biodiesel Board, "Berkeley Goes Biodiesel," press release, June 24, 2003.
23. National Biodiesel Board, "Harvard Makes Smart Move to Biodiesel," press release, February 20, 2004.
24. "Purdue Switches to Biodiesel," *The Biodiesel Bulletin*, April 4, 2004.
25. SnowJournal.com, "Cranmore Resort First in East to Use Biodiesel to Power Groomers," February 25, 2004, http://www.snowjournal.com/article766.html.
26. National Biodiesel Board, "Biodiesel-Fueled Dragster Sets Record," press release, September 18, 2002.
27. National Biodiesel Board, "Biodiesel Earns High Marks at 2003 Michelin Challenge Bibendum," press release, September 29, 2003.

Chapter 12: Looking Ahead

1. Raffaello Garofalo, telephone interview by the author, February 24, 2004.
2. Ibid.
3. Bill Ayres, telephone interview by the author, April 9, 2004.
4. Tom Leue, telephone interview by the author, April 30, 2004.
5. Joe Jobe, telephone interview by the author, May 11, 2004.
6. Daryl Reece, telephone interview by the author, November 24, 2003.

7. Mark Zappi, Rafael Hernandez, et al., *A Review of the Engineering Aspects of the Biodiesel Industry*, a report for the Mississippi Biomass Council (Jackson, Miss.: Mississippi University Consortium for the Utilization of Biomass, Mississippi State University, August 2003), 15.
8. Leue, interview.
9. Charles Peterson, telephone interview by the author, January 12, 2004.
10. Dieter Bockey, *Situation and Development Potential for the Production of Biodiesel—An International Study* (Berlin: Union zur Förderung von Oel- und Proteinpflanzen e.V., 2002).
11. Werner Körbitz, telephone interview by the author, January 12, 2004.
12. Jerrel Branson, telephone interview by the author, November 13, 2003.
13. Jon Van Gerpen, telephone interview by the author, April 12, 2004.
14. Branson, interview.
15. Dennis Griffin, telephone interview by the author, April 19, 2004.
16. Jobe, interview.
17. Jeff Probst, telephone interview by the author, May 6, 2004.
18. Leon Schumacher, telephone interview by the author, April 2, 2004.
19. Rico Cruz, telephone interview by the author, April 13, 2004.
20. Gene Gebolys, telephone interview by the author, April 20, 2004.
21. Jobe, interview.
22. Ibid.
23. Schumacher, interview.
24. Garofalo, interview.
25. Probst, interview.
26. Gebolys, interview.
27. Bob King, telephone interview by the author, May 4, 2004.
28. Gary Haer, telephone interview by the author, April 14, 2004.
29. Darryl Melrose, telephone interview by the author, March 23, 2004.
30. Gebolys, interview.
31. Reece, interview.

GLOSSARY

alkyl ester. A generic term for any alcohol-produced vegetable-oil esters or biodiesel.

aromatic. A chemical such as benzene, toluene, or xylene that normally is present in exhaust emissions from diesel engines running on petroleum diesel fuel. Aromatic compounds have strong, characteristic odors.

available production. The biodiesel-production capacity of refining facilities that are not specifically designed to produce biodiesel.

batch process. A method of making biodiesel that relies on a specific, limited amount of inputs for a single batch.

biodiesel. A clean-burning fuel made from natural, renewable sources such as new or used vegetable oil or animal fats.

biofuel. A fuel made from biomass resources, such as ethanol, methanol, or biodiesel.

bioheat. A name sometimes applied to biodiesel when it is used for heating purposes.

biomass. Plant material, including wood, vegetation, grains, or agricultural waste, used as a fuel or energy source.

bio-naphtha. A term used in some eastern European nations for biodiesel.

brown grease. The least expensive category of waste grease in the United States, produced usually from restaurant grease traps or rendering plant sludge.

Btu. British thermal unit(s), a quantitative measure of heat equivalent to the amount of heat required to raise 1 pound of water by 1 degree Fahrenheit.

carbon dioxide (CO_2). A product of combustion and a so-called greenhouse gas that traps the earth's heat and contributes to global warming.

carbon monoxide (CO). A colorless, odorless, lethal gas that is the product of the incomplete combustion of fuels.

catalyst. A substance that, without itself undergoing any permanent chemical change, facilitates or enables a reaction between other substances.

cetane number. A measure of the ignition qualities of diesel fuel.

cloud point. The point at which biodiesel fuel appears cloudy because of the formation of wax crystals due to cold temperatures.

coking. The formation of harmful carbon deposits on internal components of diesel engines.

compression-ignition engine. An engine in which the fuel is ignited by high temperature caused by extreme pressure in the cylinder, rather than by a spark from a spark plug. Diesel engines are compression-ignition engines.

continuous deglycerolization. One of a number of continuous-flow processes for making biodiesel.

continuous-flow process. A general term for any of a number of biodiesel production processes that involves the continuous addition of ingredients to produce biodiesel on a continual, round-the-clock basis, as opposed to the batch process.

dedicated production. The biodiesel-production capacity of refining facilities that are specifically designed to produce biodiesel.

diester. The French term for biodiesel, from the contraction of the words *diesel* and *ester*.

direct-injection engine. A diesel engine in which the fuel is injected directly into the cylinder. Most new diesel engines have turbo direct injection (TDI).

emissions. All substances discharged into the air during combustion.

energy crops. Crops grown specifically for their energy value.

energy efficiency ratio. A numerical figure that represents the energy stored in a fuel compared to the total energy required to produce, manufacture, transport, and distribute it.

ethanol. A colorless, flammable liquid that can be produced chemically from ethylene or biologically from the fermentation of various sugars from carbohydrates found in agricultural crops and residues from crops or wood. Also known as ethyl alcohol, alcohol, or grain spirits.

ethyl ester. Biodiesel that is made with the use of ethanol.

fatty acid methyl ester (FAME). Another term for biodiesel made with methanol.

fatty acid alkyl ester. Another term for biodiesel made from any alcohol.

feedstock. Any material converted to another form of fuel or an energy product.

flash point. The temperature at which a substance will ignite (for biodiesel, above 260°F or 126°C).

fossil fuel. An organic, energy-rich substance formed from the long-buried remains of prehistoric organic life. These fuels are considered nonrenewable, and their use contributes to air pollution and global warming.

gaseous emissions. Substances discharged into the air during combustion, typically including carbon dioxide, carbon monoxide, water vapor, and hydrocarbons.

gasohol. A fuel blend of ethanol made from fermented corn and gasoline.

gel point. The point at which a liquid fuel gels (changes to the consistency of petroleum jelly) due to extremely low temperature.

glycerin. A thick, sticky substance that is part of the chemical structure of vegetable oils and is a by-product of the transesterification process for making biodiesel. Glycerin is often used in the manufacture of soap and pharmaceuticals.

greenhouse effect. The heating of the atmosphere that results from the absorption of re-radiated solar radiation by certain gases, especially carbon dioxide and methane.

indirect-injection engine. A (typically older) diesel engine in which the fuel is injected into a prechamber, where it is partly combusted, before it enters the cylinder.

methanol. A volatile, colorless alcohol, originally derived from wood, that is often used as a racing fuel and as a solvent. Also called methyl alcohol.

methyl ester. Biodiesel that is made with the use of methanol.

multifeedstock. Used to describe a biodiesel process technology that is capable of using a wide variety of feedstock inputs.

neat biodiesel. Pure biodiesel or B100.

nitrogen oxides (NOx). A product of combustion and a contributing factor in the formation of smog and ozone.

oleochemicals. Chemicals derived from biological oils or fats.

particulate emissions. Substances discharged into the air during combustion. Typically they are fine particles such as carbonaceous soot and various organic molecules.

petrodiesel. Petroleum-based diesel fuel, usually referred to simply as diesel.

photosynthesis. A process by which plants and other organisms use light to convert carbon dioxide and water into a simple sugar. Photosynthesis provides the basic energy source for almost all organisms.

pour point. The temperature below which a fuel will not pour. The pour point for biodiesel is higher than that for petrodiesel.

products of combustion. Substances formed during combustion. The products of complete fuel combustion are carbon dioxide and water. Products of incomplete combustion can include carbon monoxide, hydrocarbons, soot, tars, and other substances.

renewable energy. An energy source that renews itself or that can be used today without diminishing future supply.

soydiesel. A term used in the United States for biodiesel made from soybean oil.

sustainable. Used to describe material or energy sources that, if carefully managed, will provide at current levels indefinitely.

transesterification. A chemical process that uses an alcohol to react with the triglycerides contained in vegetable oils and animal fats to produce biodiesel and glycerin.

triglycerides. Fats composed of three fatty-acid chains linked to a glycerol molecule.

turbo direct injection (TDI). See *direct-injection engine*.

viscosity. The ability of a liquid to flow. A high-viscosity liquid flows slowly, while a low-viscosity liquid flows quickly.

yellow grease. A term used in the United States to refer to recycled cooking oils.

BIBLIOGRAPHY

Alovert, Maria "Mark." *Biodiesel Homebrew Guide: Everything You Need to Know to Make Quality Alternative Diesel Fuel Out of Waste Restaurant Fryer Oil*. Version 9. San Francisco: self-published, May 8, 2004. Available online at www.localb100.com.

Campbell, C. J. *The Coming Oil Crisis*. Essex, England: Petroconsultants S.A., 2004.

Mittelbach, Martin, and Claudia Remschmidt. *Biodiesel: The Comprehensive Handbook*. Graz, Austria: self-published, 2004. Contact: mittelbach_biodiesel@gmx.at.

Tickell, Joshua. *From the Fryer to the Fuel Tank: The Complete Guide to Using Vegetable Oil as an Alternative Fuel*. 3rd ed. New Orleans: Joshua Tickell Media Productions, 2003.

INDEX

B

G

R

T

U

V

W

Y